Automotive Electricity

Automotive Electricity

Electric Drives

Edited by
Joseph Beretta

ISTE

WILEY

First published in 2005 France by Hermes Science/Lavoisier entitled: *Le génie électrique automobile: la traction électrique* © LAVOISIER, 2005
First published in 2010 Great Britain and the United States by ISTE Ltd and John Wiley & Sons, Inc.

ISTE Ltd
27-37 St George's Road
London SW19 4EU
UK

www.iste.co.uk

John Wiley & Sons, Inc.
111 River Street
Hoboken, NJ 07030
USA

www.wiley.com

Library of Congress Cataloging-in-Publication Data

Electronique, électricité et mécatronique automobile. English
Automotive electricity : electric drives / edited by Joseph Beretta.
p. cm.
Includes bibliographical references and index.
ISBN 978-1-84821-095-0
1. Electric automobiles--Motors. 2. Electric automobiles--Electric equipment. 3. Electric driving. I. Beretta, Joseph. II. Title.
TL220.E48 2009
629.22'93--dc22
[redacted] 2009017636

British Library Cataloguing-in-Publication Data
A CIP record for this book is available from the British Library
ISBN 978-1-84821-095-0

Printed and bound in Great Britain by CPI Antony Rowe, Chippenham and Eastbourne.

[redacted]

Table of Contents

Preface

Since the beginning of the century, electrical engineering has invaded our daily life (light bulbs, electric robots, etc.). It is present in the majority of our everyday objects.

Today it is strongly involved in the automotive market. While the change in this field has been very slow over the last ten last years, it is now beginning to accelerate and we are witnessing a wave driven by regulatory constraints and market laws which are sweeping away the last bastions of resistance.

Even if the electric car has not experienced real success, automotive electricity and electronics now hold an important place.

I dedicate this book to all of the pioneers who fought against the reservations and resistance of the system so that electrical engineering could find its place, to all those visionaries and dreamers with genius ideas, who still believe in the electric car and who are delighted by the progress of hybrid cars: for in a way, this book is also their work.

Joseph BERETTA

Chapter 1

Introduction

Throughout the history of Mankind, human beings have endeavored to extend the radius of their activities, which has always led them to improve transport techniques.

Each time new progress was made with transport, this altered humans' lives. Today, it is mobility concepts that are the focus. This mobility has multiple implications; it supports the choices made for our environment, the rules of traveling and the design of "automobiles" (cars). It is to cars themselves, and particularly to automotive electrical engineering, that we devote this work.

We will review all of the electric technologies that are used, with this first volume focusing on technologies relating to electric drive-trains.

1.1. Automotive constraints

Having come into existence more than one hundred years ago, cars are now a predominant part of our everyday lives.

Chapter written by Joseph BERETTA

It is a very original "thing", which, as the years have passed, has managed to make a place for itself as a method of transport, a high-tech object, a consumer good and a representation of our social behavior.

The future of this "thing" in the coming years is thus a captivating subject for thought.

The car, this method of preserving our individual freedom as we travel, today forms part of our daily life, and has largely surpassed its functional role; it is a symbol representing our identity and our subconscious.

Whereas in the past, during the growth phase of this market, work was primarily entrusted to engineers, today it is a process which closely associates both technical and market roles.

This goes as far as anticipating customer expectations by introducing innovations which offer new products or services corresponding to latent needs.

Success in this field will come from a subtle mixture of pragmatic vision and the mastering of technology. In this combination, electricity will play a fundamental role and will contribute to achieving the new goals of the automotive industry in terms of safety, comfort and environment.

1.2. Key figures from the automotive industry – data from the CCFA (association of French car manufacturers)

The automotive industry is a first-rank industry as a result of its significant presence.

For France, the car manufacturing sector represents 100 billion Euros in turnover, i.e. 5 to 6% of the GDP (gross domestic product), and it employs 350,000 people.

In the same vein, the entire automotive sector represents around 2.5 million jobs (including 450,000 in upstream industries, 600,000 in services related to usage and 1 million in the transportation of goods and travelers). Research and development play an important role, with 17,000 jobs and 8 billion Euros largely financed on equity.

The worldwide automotive market, which has been in constant progression since 1998, represented nearly 70.3 million PV + LCV[1] units in 2007 (4.1% growth). This growth draws more on the emerging markets (China, India, Iran, Mercosur, etc.) than on the historically large markets of North America, Western Europe and Japan.

The evolution of worldwide automotive sales since 2000 has been marked by stability, even the relative stagnation of the Western Europe and North-America markets. The Asian market, meanwhile, has grown by more than one million units each year since 2000, benefiting in particular from China's economic ascension.

The situation of the European automotive market largely reflects the economic circumstances of the various countries within the zone. In Germany, where the economy is marked by relative gloom, the automotive market has been in constant decline since 2000.

In France, the market has also registered a slight drop because of a lack of vigor in household consumption and the tendency to put money into savings, amidst a context of persistent unemployment. Another basic tendency of the European market is the regular progression of diesel motorizations: their share, on the passenger-vehicle market, rose from 24.8% in 1998 to 52.6% in 2007.

---

1 PV = passenger vehicles, LCV = light commercial vehicles (under 3.5 T).

The continent of North America is today the world's number-one zone in terms of automotive sales, with 23.8 million units in 2007. Just like Western Europe, the North-American markets (the United States and Canada along with Mexico) have presented relative sales stability.

North America is characterized by the prevalence of "light trucks", i.e. pick-ups, vans and large all-terrain vehicles. For several years the North-American market has been experiencing a major price-war between the various manufacturers involved. The "Big Three", that is, the three historically major American manufacturers, namely General Motors, Ford and DaimlerChrysler, have been suffering a constant erosion of their market shares because of the constant progression of Japanese and Korean constructors.

The Asia-Pacific zone is characterized by the sustained development of its automotive market. This "boom" reflects above all the dynamism of China, which recorded a GDP increase of more than 11% in 2007 and saw its market increase by almost 40% for several consecutive years.

Japan, the number-one market within the zone and the world's second largest automotive market, is characterized by sales stability. The Japanese market is very slowly opening up to foreign automotive imports.

Chapter 2

Basic Definitions

2.1. Basic concepts

2.1.1. *Basics of automotive energy*

Most of the energy introduced into a vehicle is lost during transfers (friction, heat, pumping). Manufacturers continue to explore a number of possibilities for reducing these losses.

To talk about energetic concepts, we need to talk about efficiency.

Efficiency is the ratio of energy used with respect to the work involved in setting the vehicle in motion. It directly affects the consumption: the greater the efficiency, the lower the fuel consumption of the car.

– *Let us examine how energy in a car is reduced.*

When energy is introduced into an engine, only 30% remains when it comes to setting the wheels in motion. There are, throughout the process, losses which lower the efficiency. We estimate that 30% of energy is lost in the

---

Chapter written by Joseph BERETTA

form of heat from the engine, approximately 30% leaves in the exhaust gas and 10% is dissipated by mechanical friction and driving the accessories (water pump, air-conditioning, etc.).

On arrival, the remaining 30% are reduced slightly further by the mechanical efficiency of the gear box and the transmissions.

Some of these losses are used to provide other services: the heat released by the cooling system is thus used for heating the cabin, the heat released through the exhaust supports the post-treatment mechanisms.

– *Each transformation has its own efficiency.*

The total efficiency of an engine (equal to 0.3 in the best cases) is the relationship between the energy supplied to the crankshaft and the energy supplied by the fuel. More precisely, it is the result of the product of two outputs:

1) The efficiency of the chemical reaction, which breaks down into:

– theoretical thermodynamic efficiency of the driving cycle, which depends on the compression ratio;

– efficiency with the additional losses, which expresses the actual reduction compared to the theoretical reduction (inertia, viscosity, pumping, etc.);

– efficiency of combustion (combustion does not use all the energy supplied by the fuel).

2) The mechanical efficiency, which comes from friction in the moving parts in the engine and from the work dissipated to drive all the accessories (water pump, injection, air-conditioning, etc.).

– *Efficiency varies according to the type of engine (gasoline or diesel).*

Theoretically, the gasoline engine offers better efficiency in terms of thermodynamics. However, diesel presents a more favorable result overall, because of its higher volumetric ratio (approximately 18/1 compared with 10/1 for gasoline) and a low level of losses by pumping.

– *How to improve the efficiency:*

– increase the volumetric[1] ratio of compression of gasoline engines, in order to improve the thermodynamic efficiency;

– reduce the losses (thermal or by pumping);

– optimize the shape of the combustion chambers, their internal aerodynamics in particular, in order to optimize combustion. In the next few years, the efficiency could increase by 10 to 20%, thanks to the conjunction of various technologies: gasoline direct injection, "downsizing"[2] of engines associated with strong turbo-booster and variable distribution.

2.1.2. *Basics of automotive dynamics*

2.1.2.1. *Useful reminder of automotive dynamics*

The force opposing the car's displacement can be calculated as follows:

$$Fw = Fro + Fl + Fst$$

---

1 The volume ratio indicates the ratio of the volumes remaining above the piston, between the base position (bottom dead center) and upper position (top dead center) of the piston. It directly influences the thermodynamic output of the cycle.

2 Tendency which consists (to lower consumption) of reducing the swept volume of an engine, while preserving the same performance level. This is obtained by supercharging, which makes it possible to obtain strong specific performance (power with respect to the swept volume of the engine).

with:

Fw = resistance to movement;

Fro = rolling friction;

Fst = slope resistance;

Fl = air resistance.

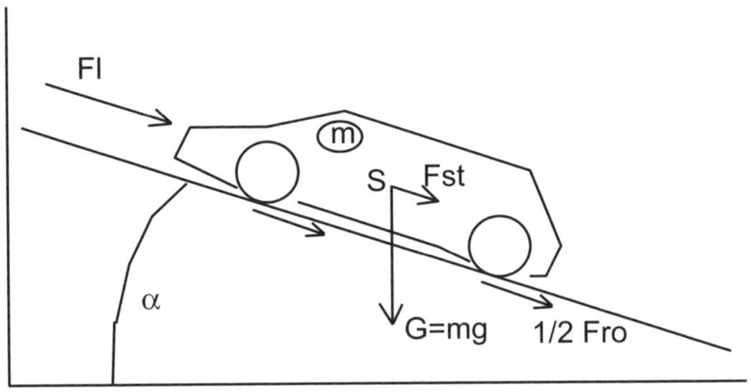

The various forces can be calculated:

Fro = *f*.m.g with f = rolling friction coefficient ≅ 0.025;

Fl = 0.5 x ρ x Scx (v + vo)² with ρ = 1.2 kgm³ and Scx ≅ 0.3 m²;

Fst = mg sin α ≅ mg p%;

Fw ≈ *f* mg + mg p% + fl;

SCx= (coefficient of drag x front surface).

2.1.2.2. *The drive force*

Diagram of a motor reducer-wheel assembly.

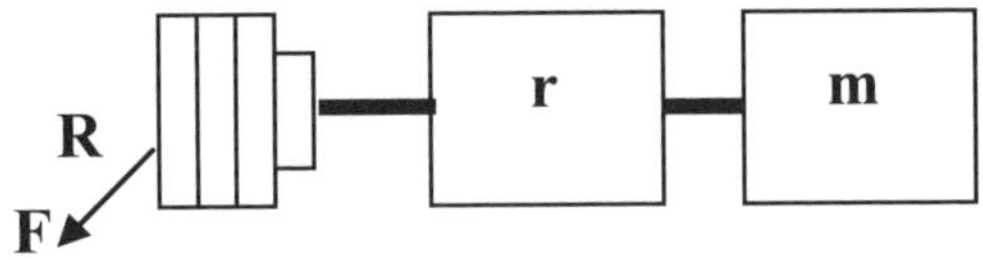

The drive force F is expressed as follows:

$$F = \frac{Cm.\ r.\ \eta}{R}$$

F = Drive force;

R = Radius of the wheel;

m = Electric motor;

r = Ratio of transmission Vm/Vr, where:

Vm = engine speed,

Vr = wheel speed;

Cm = engine torque;

η = transmission efficiency ≈ 0.90 to 0.98.

If F > Fw ➔ then the vehicle is in acceleration.

If F < Fw ➔ then the vehicle is in deceleration.

We can easily calculate the power at the wheels:

Pw = Fw . v

where:

Pw = Power at wheels;

v = Vehicle speed;

Fw = Resistance to movement.

If we look at the characteristics of the electric motor (torque/speed):

Ωr = Speed of rotation of the wheel;

Cr = Torque at the wheel;

R = Radius of the wheel;

η = Efficiency of gear box;

r = Ratio of gear box;

Ωm = Engine rotation speed;

Cm = Engine torque.

The conditions to be met in order to define the main characteristics of the electric motor "m" are:

Cr defines the hill start Cr = F . R;

Ωr defines the maximum speed $\Omega r = V/R$.

This enables us to define the characteristics of the engine (Ωm and Cm).

2.2. The different electric drive-train systems

2.2.1. *Basic definitions*

Energy

Value characterizing a system and expressing its capacity to modify the state of other systems.

Nature of the energy:

Characterizes the various forms which energy can take (mechanical, electric, chemical, hydraulic, thermal, radiant or nuclear).

Energy type-definition

Characterizes energy parameters of the same nature (for electric energy: voltage, current, frequency).

Energy chain

Characterizes the nature of the energy substance that may be transferred between the infrastructure and the vehicle (gasoline, gas, electricity, etc.).

Energy storage device

System allowing energy to be stored without modifying the nature and type definition of the flow of energy (input and output).

Primary energy source

Set of raw materials or natural phenomena used for energy production.

Generator of energy or energy source

System allowing the production (generator) or the delivery (source) of energy, starting from a primary energy source (raw materials: hydrocarbons, coal, uranium, etc., or natural phenomena: wind, sun, gravity, etc.).

Generator or source of energy on-board a vehicle

System allowing the production and/or the delivery of energy for its use in the vehicle; associated with an energy carrier, it is made up of a storage system as a minimum.

Transmitter of energy

System which retains the nature of the energy but changes its type definition (gearbox, AC/DC electric converter).

Energy converter

System which changes the nature of energy (engine, radiator, etc.).

Single-energy-chain vehicle

Vehicle in which supply takes place through only one energy chain.

Multi-energy-chain vehicle

Vehicle in which supply takes place through more than one energy chain.

Single-energy-system vehicle

Vehicle using only one energy generator and only one energy transformer for its propulsion.

Multi-energy-system vehicle

Vehicle using more than one energy generator and at least one energy transformer for its propulsion

Hybrid vehicle

Multi-energy-system vehicle where at least one energy generator is reversible.

Non-hybrid vehicle

Multi-energy-system vehicle where all the energy generators are irreversible.

Energy pathway

This is the path taken by energy from an energy source or storage unit towards the wheels or another storage unit.

Single-mode vehicle

Vehicle where the selection of the modes (association of the energy pathways, allowing the drive force to be delivered) is carried out in an automatic manner by the management system of the vehicle.

Multi-mode vehicle

Vehicle where the selection of the modes (association of the energy pathways allowing the drive force to be delivered) can be imposed by the user or a system external to the vehicle.

The tree structure of these systems can be presented as follows, whereby we can distinguish various levels: infrastructure, vehicle, drive-train system and components.

The infrastructure level defines the connections between the vehicle and this infrastructure (number and type of connection, at this infrastructure).

The vehicle level defines how the various energy systems that contribute to propulsion (hydrid system, operating procedure for these systems, etc.) are used.

The drive-train system level defines how the drive force of the vehicle is generated.

Of course, these levels are interlinked, and due to language misuse we often mistake the level when we talk of hybrid.

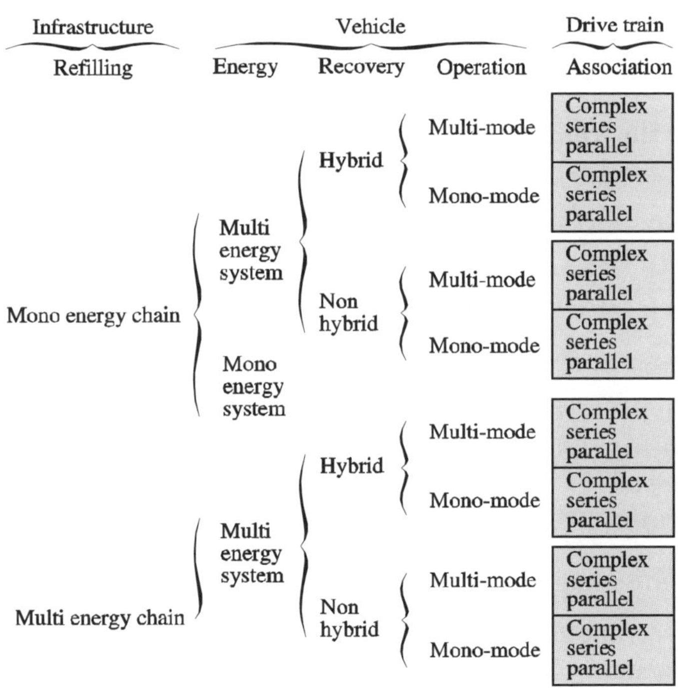

Some examples:

Electric-thermal hybrid vehicles belong to multi-energy systems, and to the multi-energy-chain category if they can be both recharged on the electrical network, and have their the tank filled with hydrocarbon fuels.

The power-assist system cannot be recharged on the electricity network, so it belongs to the single-energy-chain category.

Thermal and electric vehicles are single-energy chains and single-energy systems.

Dual-fuel thermal vehicles are multi-energy chain/multi-energy system/non-hybrid/dual-mode/series.

Fuel-cell vehicles using stored oxygen and hydrogen are multi-energy chain/multi-energy system/non-hybrid/single-mode/series. If a battery is added they become: multi-energy chain/multi-energy system/hybrid/complex series.

In the remainder of the chapter, we will focus solely on the drive-train systems level and will detail the components level.

2.2.2. *Definitions of drive-train systems*

Drive-train system

An assembly of components crossed by the energy flow which provide a vehicle with its capacity for movement. It is composed of a traction system and an on-board energy generator.

Traction system

A component of the drive-train system providing the mechanical transmission of movement. It is composed of the wheels and its differential, the transmission or gear box and an engine converting energy provided from the generator into mechanical energy.

Generator or on-board energy source

A component of the drive-train system ensuring the storage and conversion of energy; it is composed of a storage system and a conversion system (converter and/or transmitter).

It is possible for there not to be a conversion system when the engine directly accepts the nature of the energy stored in the tank.

These definitions make it possible to construct the diagram of a drive-train system.

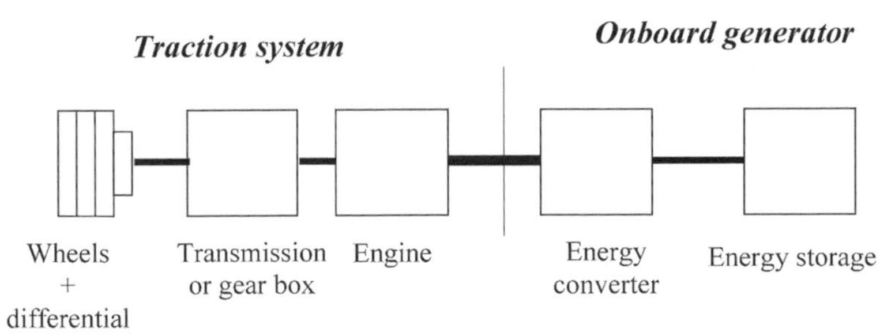

Figure 2.1. *Drive-train system*

Now that we have these definitions and a representation of the drive-train systems, we can move on to hybrid systems.

Hybrid drive-train system

This is a drive-train system created through the hybridization of two or more single drive-train systems.

Based on the previous definitions, let us now define series and parallel hybrid drive-train systems

Series hybrid drive-train system:

– it is an energy-generator hybrid;

– the transmission of the movement is performed by a single engine only;

– the on-board energy generator is thus created through the hybridization of two or more energy generators.

Parallel hybrid drive-train system:

– it is a traction-system hybrid;

– the traction system is created through the hybridization of two or more traction systems;

– it is of course necessary to associate with each traction system a suitable energy generator;

– the transmission of the movement is assured by several engines.

However, to carry out all of these changes, it is necessary to introduce the concept of a coupling component.

Coupling component

This is a hybrid drive-train system component which makes it possible to connect the single drive-train systems making up the hybrid.

To finish the breakdown of these definitions, we will now examine the sub-categories of hybrid drive-train systems.

Thus, if we consider the standard diagram of a drive-train system, we can imagine that each component is transformed into a coupling component.

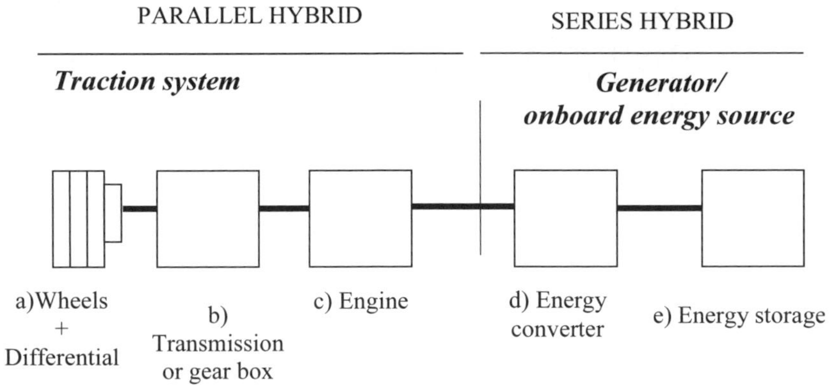

Figure 2.2. *Types of hybrid*

a) If the coupling component is positioned at the wheels we talk of a *double drive-train system parallel hybrid* and the diagram can be broken down as follows.

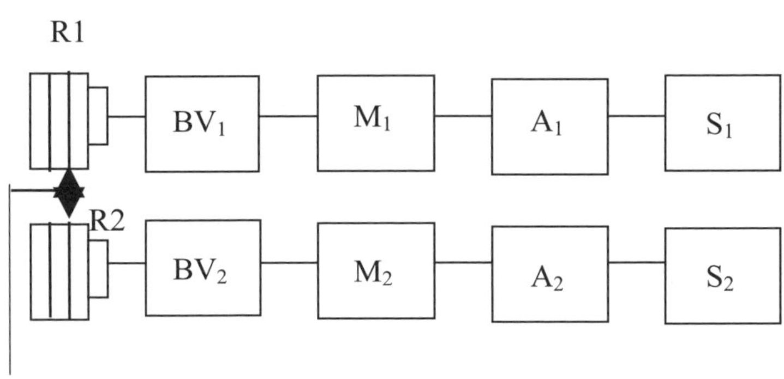

Figure 2.3. *Double drive-train system parallel hybrid*

The road and the wheels produce the coupling component.

b) If the coupling component is positioned at the gear box, we talk of a *double-shaft parallel hybrid* and the diagram can be broken down as follows:

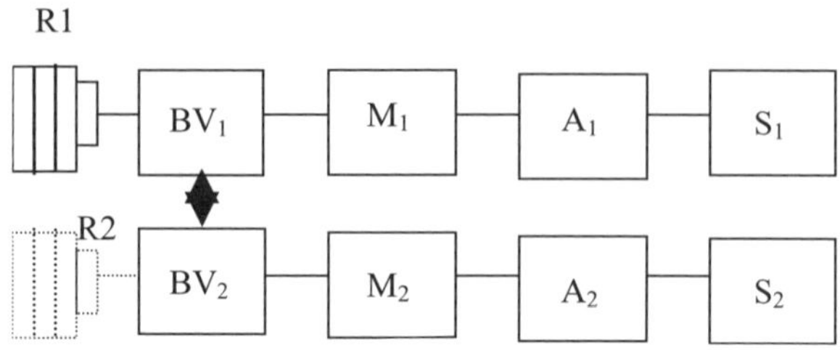

Figure 2.4. *Double-shaft parallel hybrid*

The speed ratios between M1 and M2 are not fixed.

c) If the coupling component is positioned at the engine, we talk of a *single-shaft parallel hybrid.*

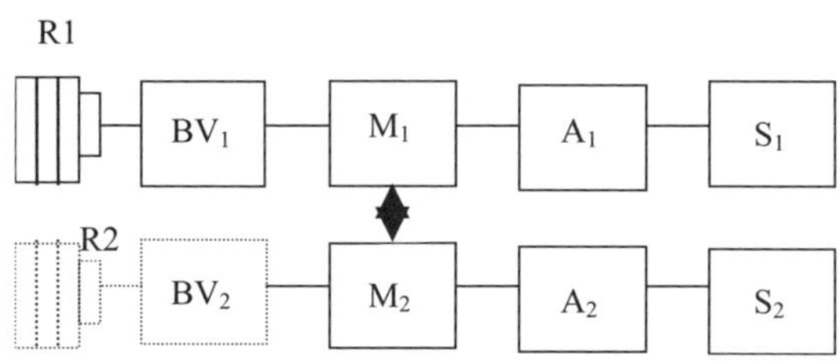

Figure 2.5. *Single-shaft parallel hybrid*

The engines M1 and M2 turn with fixed speed ratios.

d) If the coupling component is positioned at the energy converter system, we talk of a *double-energy-generation series hybrid*; through misuse of the series hybrid.

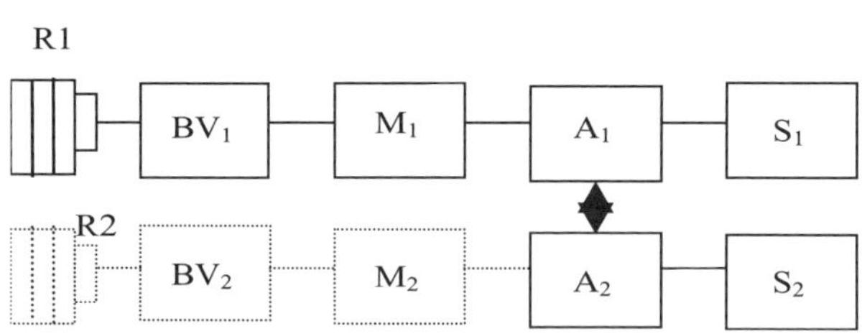

Figure 2.6. *Double-energy-generation series hybrid*

e) If the coupling component is positioned at the storage system we talk of a *double-energy-storage series hybrid.*

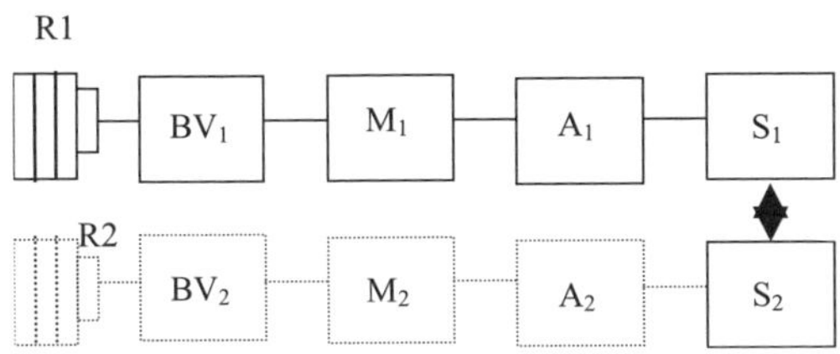

Figure 2.7. *Double-energy-storage series hybrid*

– To finish, set out below is the concept of hybridization ratio:

$$\text{Parallel hybrid ratio Phr} = \frac{PM2}{PM1 + PM2}$$

$$\text{Series hybrid ratio Shr} = \frac{PS2}{PM1}$$

PM1 = Engine power on drive-train 1.

PM2 = Engine power on drive-train 2.

PS2 = Generator power on drive-train 2.

This entire approach is generic and entirely independent of hybrid-type systems: thermal, electric, hydraulic or others.

2.2.3. *Thermal-electric hybrid systems*

We will now limit our focus to electric thermal systems, imposing certain parameters.

For parallel:

M2 = internal combustion engine ICE;

M1 = electric motor EM;

S1 = electric storage ES.

For series: S2 = fuel tank T:

M1 = electric motor EM;

S2 = electric generator EG;

S1 = electric storage ES.

A1 & A2 are converter systems between energy storage and the engine.

R1 & R2 are wheels and we obtain the following standard diagrams

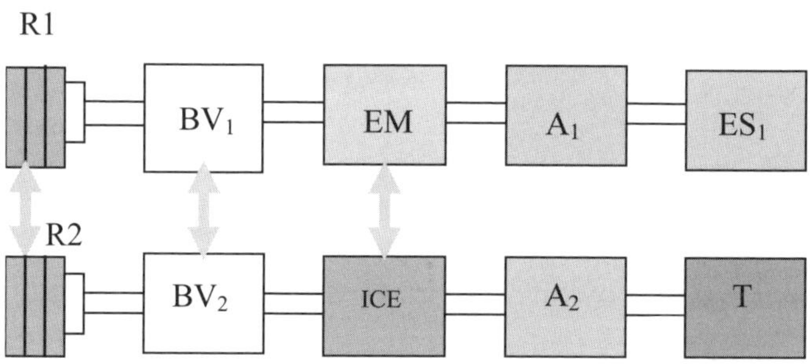

Figure 2.8. *Thermal-electric parallel hybrid*

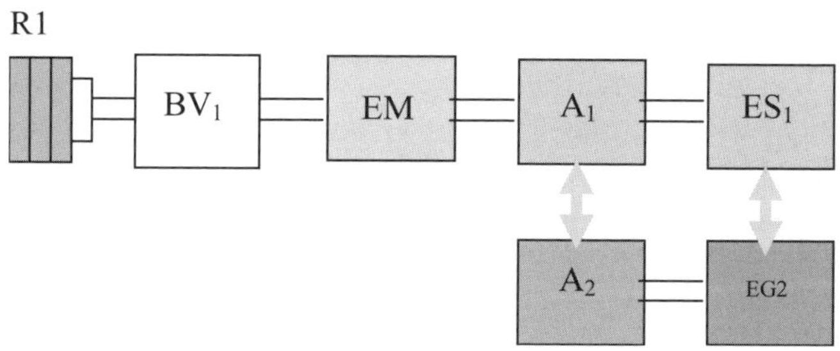

Figure 2.9. *Thermal-electric series hybrid*

The hybridization ratio becomes: Phr = Pice/(Pice + Pem); SHr = Peg/Pem, where:

Pice = Power of the heat engine;

Pem = Power of the electric motor;

Peg = Power of the electric generator.

2.2.3.1. *Fuel-cell systems*

Fuel-cell systems can be broken down as follows:

On-board generator/energy source of fuel cell system.

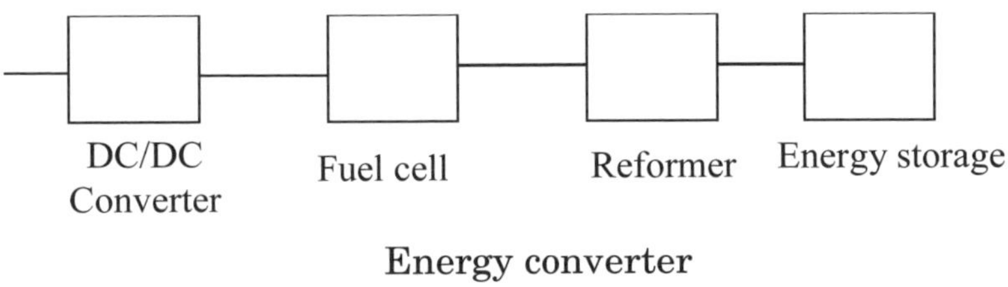

Figure 2.10. *Fuel-cell energy generator*

2.2.3.2. *Panorama of simple thermal-electric hybrids*

With of all these definitions, we can draw a table representing the panorama of simple thermal-electric hybrids.

This diagram will break down in terms of power the various possibilities of hybridization whilst gradually varying the ratio of hybridization. This very instructive diagram makes it possible to have an initial idea of the best hybridization solutions based on criteria such as efficiency, weight, cost and industrial synergy.

However, the theory does not stop there because complexity will appear when we pass from simple hybrids to complex hybrids.

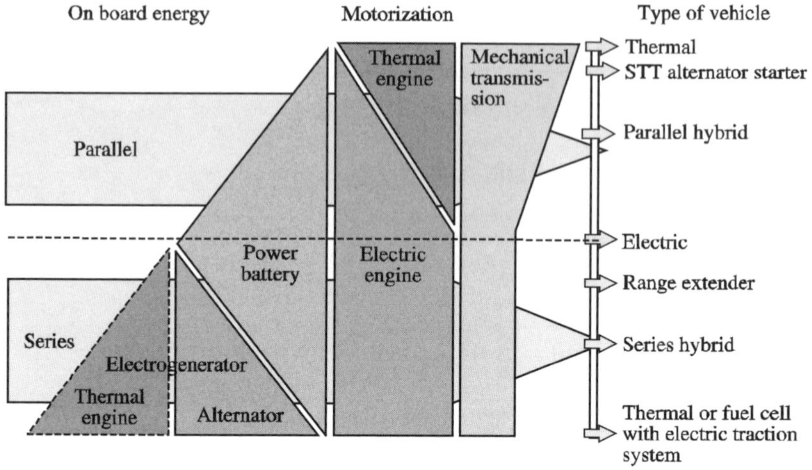

Figure 2.11. *Hybrids*

2.2.4. *Complex hybrids*

In section 2.2 we envisaged hybrids produced from the association of two drive-train systems. But why not consider the association of several systems? Thus, if we consider the standard diagram of a drive-train system we can connect up three or more systems using coupling components. In order to pursue the structuring of this assembly, it is necessary to define certain relevant indicators.

The order of the drive-train system "O": this is the number of simple system, that needed to be associated to create the final system.

The index of the drive-train system "I": this is the number of coupling components contained in the final system.

The degree of performance of the drive-train system "DP": this is the sum of the multiplication of the efficiency of the components crossed for each energy pathway, and those multiplied by the fraction of energy crossing through them.

We can thus define the following relationships between these indicators:

O-1 < I < O

Number of energy pathways = 2^{order}.

Degree of performance = $\sum$ (degree of performance of the energy pathways).

= $\sum$ (% power of the pathway x output of the pathway components) .

No. of drive-train systems	**System order**	**System index**	**Description**
1	1	0	Simple system
2	2	1	simple hybrid, parallel or series
2	2	2	simple hybrid, parallel + series
3	3	2	double hybrid, parallel or series
3	3	3	double hybrid, parallel + series
4	4	3	triple hybrid, parallel or series
4	4	4	triple hybrid, parallel+ series
N	N	n-1	n hybrid, parallel or series
N	N	n	n-th hybrid, parallel + series

To illustrate these remarks, set out below are some examples of complex hybrids.

This diagram represents the TOYOTA PRIUS system, where the coupling component consists of an epicyclical gear.

Parallel Hybrid Order 3, Index 3: Triple-Shaft Hybrid, Parallel + Series.

TOYOTA PRIUS

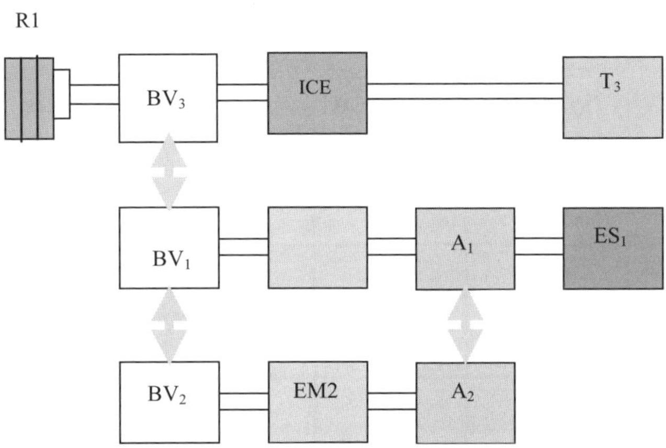

Figure 2.12. *Toyota Prius*

The coupling components BV1, BV2, BV3 represent the gear box of the TOYOTA PRIUS:

BV3 + BV1 = epicyclical gear;

BV2 = output transmission;

EM1 = electric generator;

EM2 = electric motor.

A1 + A2 is the coupling of two converters.

ES1 is the battery.

Parallel Hybrid Order 2, Index 1: Single-Shaft Hybrid.

CITROEN XSARA DYNALTO

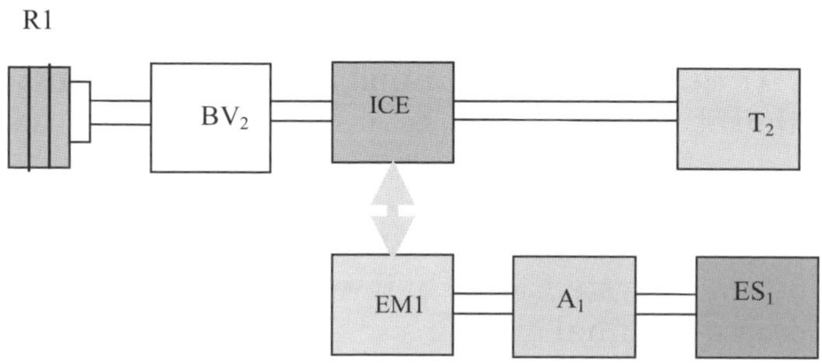

Figure 2.13. *Citroen Xsara Dynalto*

The engines EM1 and ICE turn with fixed speed ratios.

The bases of complex hybrids having now been given, it is necessary to define the peripheral parameters.

The complex ratio of hybridization

As the ratio of hybridization is a ratio between an electric output and the total power, it is by nature representative of the participation of the electric chain in the longitudinal vehicle dynamic characteristics; the difficulty arises when we touch upon complex hybrids. For the two main categories, parallel and series, while there is no problem concerning simple hybrids; when it comes to complex hybrids, it is necessary to define the complex rate, which will be a multiplication of the sums of the parallel ratio and series ratio.

Complex hybridization ratio Chr = (Σparallel Rh series Rh)

The operator * is not a simple product but a complex product that takes account of the system configuration and its operating procedure. At this stage it is necessary to return to the vehicle to evaluate the ratio of hybridization. Thus, we now have three indicators characterizing drive-train systems:

1. *The Order*: this is the number of simple drive-train systems that it was necessary to associate to create the final system.

2. *The Index*: this is the number of connection components connecting each system which it was necessary to put in place to create the final system.

3. *The degree of freedom*: this is the number of energy pathways in the system.

4. *The degree of performance*: this is the sum of the multiplication of the outputs of the components crossed for each energy pathway, taking account of the fraction of energy passing through them.

Chapter 3

Electric-Powered Vehicles

3.1. History

In 1901, in view of its performance, the future looked bright for the electric vehicle. It was possible to imagine installing charging stations where during the night it would be possible to recharge a battery that was discharged after a day's use, or change an empty battery for a full one. However from 1907, the newspapers began to declare that the electric vehicle was in decline: although it did not present any disadvantages at start-up and it was clean, it remained a city car or one of luxury, very easy to drive, for example, on a small excursion. It only had one electric motor, light but robust: on slopes, for example, it could develop power twice that of normal power without dangerous overheating. Its two 110 V battery groups (at the front and back), were high-capacity and robust accumulators for long journeys. But its price still often remained rather high, due to the accumulator battery itself, which provided its autonomy. In fact, the conclusion was quickly reached that: *"The electric car has hardly progressed in the last ten years, and we can*

Chapter written by Joseph BERETTA, Cyriacus BLEIJS, François BADIN and Thierry ALLEAU.

say that, generally speaking, it is not practical. We have managed to develop cars able to cover nearly 100 km without recharging, and I believe that it is a maximum. Moreover, the speeds reached are low because we are obliged to protect the accumulators, and these are, in addition, always heavy and cumbersome. The electric car can, in special circumstances (proximity of an electric factory, a defined, regular journey), be of some use, but this remains modest. The considerable weight of the accumulators always causes strong wear of the tires. The advantages, in the final analysis, are largely outweighed by the disadvantages".

Lastly, the popularization of the 1909 Ford T model triggered a popularization of vehicles powered by gasoline, a far superior material in terms of being an autonomous energy source, and this was the start of their durable influence on the market, all the more so as electric vehicles were no longer easily accepted because of reservations regarding their cost and their performance (speed and range). At the time of World War I, electric vehicles fell very much to the wayside.

In the United States, long-lasting success:

In the new continent, the adventure of the electric car began in 1894 in Philadelphia where two manufacturers, Henry Morris and Pedro Salom, manufactured a first experimental car: the *Electrobat*. It was a vehicle equipped with an electric motor for each wheel and whose autonomy was 40 km at a constant speed of 32 kph. As of the following year, they began mass-producing it, in particular for Philadelphia and New York taxi fleets.

But it was with the industrialist Albert Pope (cycle manufacturer) that the market really took off. He proposed a whole range of electric vehicles, in particular the *Columbia*, a city car with two or four seats, which was a great success. Then came to be added to the market already famous brands like *Baker* of Cleveland, *Riker* of Elizabethport and *Wood* of Chicago. The annual rate of production of electric vehicles was then approximately 500 units. *Detroit Electric* joined their ranks in 1907 and became the best-known brand, experiencing the greatest longevity, until 1942. The *Detroit* car could reach 36 kph. It was equipped with a lead-acid battery composed of 42 cells and 15 plates, with a capacity of 185 Ah. They had five speeds going from 9 to 36 kph. Steering controls at the disposal of its driver were of a remarkable simplicity and as low as possible in number. That allowed a weight reduction, fundamental for the electric car, for which reducing it is equivalent to increasing the operating range. The brake command (a small pedal) acted on the wheels and on the electric current which could be cut instantaneously by pressing on the pedal. The only complicated component was the controller: "*As the electric car has no gearbox, gear shifting is via this specific component, with which we regulate the engine speed and consequently the car speed. The 'controller' plays, all in all, the role of the rheostat*"[1].

American electric cars were often equipped with Edison accumulators (nickel/iron accumulators). They presented many advantages in relation to lead-acid accumulators, like being able to be recharged with a high current, therefore in a much faster way, and being used until complete discharge of the battery. However, they presented the disadvantage of a dangerous hydrogen release during the charge. Lead-acid batteries improved and their cyclability performance was reinforced. But although they did not cease to improve, moving from 13 to 18 Wh/kg between 1913 and 1930, maintenance of the vehicle was expensive, which made it lose a part of its appeal[2,3].

---

1 WOLF R., *Le Véhicule Electrique gagne le Cœur de la Ville*, Centre Français de l'Electricité, 1995.

2 *La Vie Automobile*, no. 150, 13 August 1906, p. 524.

At that time, in the United States, the electric vehicle was witnessing very strong interest from city-based female customers. Indeed, in 1914, the merits of the electric vehicle were described in a Detroit Electric advertisement: *"You stop at a crossroads. You are surprised; the car in front of you is driven by a woman! ... Detroit Electric has put the control of a car in the hands of ladies"*. Indeed, battery-powered cars avoided needing to use the crank to start-up. Pleasant control and silence were two other assets for these customers not at all put off by the limited operating range and the low speed of the vehicle for driving in town or fashionable visits.

The (very fleeting!) success of electric vehicles was such that B.S. Hender estimates that at the beginning of the century, there were several tens of thousands in circulation throughout the world. J.L. Hartman, E.J. Cairns and E.H. Hietbrink put at 10,000 the number of electric vehicles (6,000 private cars and 4,000 commercial cars) produced in the United States in 1912, the year the electric vehicle was at its peak. Two years later, the production of private cars had fallen to less than 5,000 and now accounted for only 1% of the total production of the United States, even though it was in 1914 that *Milburn Wagon Co*, an Ohio horse-drawn-carriages industry, presented an electric vehicle, of which 7,000 units were to be sold.

	Electric	**Gasoline**	**Vapor**
1899	1,575	936	1,681
1904	1,495	18,699	1,568
1909	3,826	120,393	2,374
1914	4,669	564,385	-
1924	391	3,185,490	-

Evolution of cars sold in USA according to three different energies

---

3 NICOLON A., *Le Véhicule Électrique Mythe ou Réalité?*, Editions de la Maison des Sciences de l'Homme, Paris, 1984.

As we see in the table above, sales declined to the point of practically ceasing in 1918. Indeed, different progress supported the development of gasoline cars: a higher autonomy, a practical supply, road improvements facilitating excursions out of towns, and especially, from 1912, the introduction of an associated electric starter and lighting system proposed by *Dayton Engineering Laboratories Company (D.E.L.C.O.),* signaled the end of the electric car. In the United States, in 1921, there were only 18,200 electric vehicles amongst the 9 million vehicles. Ten years later, they no longer appeared in the statistics.

3.2. Battery-powered electric vehicles

In an electric vehicle, the battery is the most cumbersome and heaviest component, that which determines the dynamic performance of the vehicle.

By electrifying a thermal vehicle, we must install a battery whose weight is compatible with the original structure, which leads to a payload, acceleration and autonomy performance that is acceptable for urban driving.

To obtain this result it is necessary to develop a calculation method for the optimization of this battery weight in order to obtain the same performance on a specific vehicle while reducing the weight and size of the electric components.

3.2.1. *Battery sizing*

The battery is the electric energy reserve. It is currently made up of several basic units.

Specific energy

Specific energy characterizes the quantity of energy which one kilogram of battery can restore. It is expressed in Wh/kg.

Various types of batteries can currently be used on vehicles:

– the lead-acid battery (Pb) whose specific energy is 27 Wh/kg.

As an example it was used on the electric Citroen C15; the battery was 500 kg and the on-board energy was 27 Wh/kg x 500 kg, that is, 13,500 Wh:

– the nickel cadmium battery (Ni-Cd) whose specific energy is 48 Wh/kg, that is, practically double that of Pb. The electric Peugeot 106 and Citroën Saxo have 260 kg of NI-Cd on-board battery, the on-board energy is 12,480 Wh;

– the nickel-metal hydride battery (Ni-MH) whose specific energy is 75 Wh/kg. It is used on Japanese and American vehicles;

– the lithium-ion battery (Li-ion) which is only just being marketed today and whose specific energy is about 120 Wh/kg. It was tested on the 106 VEDELIC prototype which has 25,000 Wh of on-board energy for 250 kg of battery. This battery is now used on all new electric vehicle projects.

Specific power

In order to accelerate the vehicle, it is necessary to supply the electric motor with a substantial quantity of energy for a few dozen seconds, i.e. to have a specific power which is the second criterion of a battery. This specific power is expressed in Wh/kg.

3.2.2. *Vehicle specifications*

The aim of this chapter is to specify the key objectives and conditions of an electric vehicle, essential for carrying out the first estimates of the battery weight and total weight of the vehicle.

There are three of them:

– the payload: this load is indicated by Mu and is expressed in kg; the higher the "objective" payload, the greater battery will be;

– the acceleration capacity: this capacity is characterized by the value in seconds necessary to accelerate the vehicle from 0 → 50 kph. The greater the desired capacity, the more powerful the battery will have to be, and the higher its weight will be;

– the urban autonomy: this autonomy "A" is expressed in km. It depends directly:

- on the energy stored in the battery;

- on the consumption "c" of the vehicle at the battery output; this consumption expressed in Wh/km/kg, corresponds to the quantity of energy necessary to cover, in urban driving conditions, one kilometer with one kilogram.

The measurements carried out on various types of urban electric vehicles indicate an urban consumption of 0.11 Wh/km/kg.

Thus, 11 kW/h is needed to ensure an urban autonomy of 100 km for a 1,000 kg vehicle (0.11 Wh/km/kg 0.100 km . 1,000 kg).

3.2.3. *Calculating the vehicle weights*

It is advisable to determine the total weight of the vehicle in functioning order so as to check at the start of the study if the project will enable the objectives laid down in the specifications to be met.

This total weight is broken down into seven basic weights:

M = Mm + Ms + Mmt + Meq + Max + Mb + Mu (1) with:

M: total weight;

Mm: mechanical weight;

Ms: weight of structure;

Mmt: weight of drive-train;

Meq: weight of equipment;

Max: weight of auxiliaries;

Mb: battery weight;

Mu: useful weight.

Mechanical weights: these include the suspension, the brakes, and the wheels. They are proportional to the total weight M of the vehicle.

$$Mm = K1 \cdot M$$

K1 mechanical weight coefficient

Weight of the structure, including the body-in-white and opening panels. It is also proportional to the total weight M of the vehicle:

$$Ms = K2 \cdot M$$

K2 structure weight coefficient.

Weight of the drive train, including: the electric motor, the reducer and the control unit.

$$Mmt = (40 + K3) . (M/250)$$

K3 drive-train weight coefficient

40 kg corresponds to the non-active weights of the motor (control and casings). 250 W/kg corresponds to the specific power of the motors, K3 is an accelerative force characteristic coefficient.

Weight of the equipment: this weight depends on the useful weight Mu. It is the sum of a contractual weight equal to 50 kg, which corresponds to the windshield, the bumpers and the rear window and a weight valued at 1/7 (14%) of the useful weight (side window panels, seats, etc.).

$$Meq = 50 + Mu/7$$

Weight of the auxiliaries: it is assumed to be equal to 30 kg:

$$Max = 30$$

The battery weight depends on the total weight of the vehicle M, on the desired urban autonomy, "A", in km, on the urban consumption of the vehicle "c" in Wh/km/kg and on the battery's specific energy "e" in Wh/kg.

$$Mb = c . A . (M/e)$$

This battery weight must satisfy one condition: to ensure that the power delivered by the battery is sufficient to accelerate the vehicle (including the driver) in accordance with the objective laid down by the specifications. It is thus advisable to check the inequality:

$$Mb > (M_{ODM} + 70).(20/p) \qquad (2)$$ with:

$(M_{ODM} + 70)$ being the weight of the vehicle in functioning order, increased by the driver's weight;

20/p being the power necessary for a correct acceleration capacity;

P being the specific power of the battery in Wh/kg.

The total weight of the vehicle in functioning order (1) is thus expressed according to the "objective" payload and the specific energy of the battery:

$$M = \frac{840 + (Mu.8)}{7((1 - k_1 - k_2 - k_3 /250).\ c.(A/e)}$$

The coefficients k_1, k_2, k_3 are different for PV (passenger vehicles) and LCV (light commercial vehicles).

For a small urban vehicle, k_1 and k_2 are given based on the results obtained on the ECO 2000 project (very low consumption vehicle in the year 2000). We retained:

$$k_1 = 0.12 \rightarrow 0.18$$

$$k_2 = 0.17 \rightarrow 0.2$$

k_3 is worth 16 watts. It is the energy restored by the engine which makes it possible to accelerate 1 kg from 0-50 kph in 7 seconds. These 16 watts correspond to 20 watts at the battery output, taking into account the motor efficiency.

The weight M of the vehicle thus has as an expression:

$$M = \frac{840 + (Mu.8)}{4.522 - 7(\frac{c}{e}.A)} \quad (3)$$

We can also quote the following relations which bind the battery weight Mb to the autonomy and the specific power of the electric motor.

Mhb = weight without battery = weight of motor + mechanical weight + weight of structure + weight of equipment + weight of auxiliaries + useful weight

$$Mhb = Mmt + Mm + Ms + Meq + Max + Mu$$

$$Mhb = M - Mb = k_1.Mb + k_2$$

$$Mb = \frac{k_3.A}{e - (k_4.A)} \quad \text{autonomy criterion}$$

$$Mb = \frac{k_5.P}{p - k_6.P} \quad \text{acceleration criterion}$$

where k_1, k_2, k_3, k_4, k_5 and k_6 are constants depending on the size of the vehicle:

e = specific energy of the battery in Wh/kg;

p = specific power of the battery in W/kg;

P = specific power of the electric motor in W/kg.

3.2.4. *Application on a small vehicle*

The preceding expression (3) makes it possible to quickly give a first order of magnitude of the vehicle weight for which the specifications envisage:

– payload: 250 kg;

– urban autonomy: 90 km;

– battery (specific energy: 48 Wh/kg).

The total weight is worth M = 923 kg which breaks up into:

– useful weight: 250 kg (fixed objective);

– mechanical weight: 111 kg;

– weight of structure: 157 kg;

– weight of drive-train: 99 kg;

– weight of equipment: 86 kg;

– weight of auxiliary: 30 kg;

– battery weight: 190 kg.

The battery is sufficient to accelerate the vehicle correctly since it is higher than the weight of the second member of this inequality (2):

$$(923 - 250 + 70)\ 20/90 = 165\ \text{kg}$$

This fast method of evaluating the vehicle weight and the various other weights makes it possible to determine the influence on the performance of modifying one or other of the objectives envisaged in the specifications or the characteristics of the battery considered.

Influence of the autonomy factor

Let us suppose that it is planned in the specifications to carry the autonomy to 120 km, that is, 30 km more. The calculations give new weights, presented in the table below.

Range	120 kph	90 kph	Objective
useful weight	250 kg	250 kg	
mechanical weight	131 kg	111 kg	
weight of structure	186 kg	157 kg	
weight of drive-train	110 kg	99 kg	
weight of equipment	86 kg	86 kg	
weight of auxiliaries	30 kg	30 kg	
battery weight	301 kg	190 kg	+ 58%
total weight	1,094 kg	923 kg	+ 19%

It is easy to check that the battery is higher than the minimum weight, which guarantees an acceleration capacity that is sufficient for such a vehicle:

$$Mb = 301 > (1{,}094 - 250 + 70)\ 20/90 = 203$$

Influence of the battery type

Let us suppose that it is envisaged in the specifications to use a higher-energy future battery, for example a battery having a specific energy of 70 Wh/kg, the specific power of which would be the same, that is, 90 Wh/kg. With these new characteristics, we obtain the weights presented in the table below.

Battery specific energy Specific power	**70 Wh/kg 90 Wh/kg**	**48 Wh/kg 90 Wh/kg**	**Target**
useful weight	250 kg	250 kg	
mechanical weight	96 kg	111 kg	
weight of structure	137 kg	157 kg	
weight of drive-train	91 kg	99 kg	
weight of equipment	86 kg	86 kg	
weight of auxiliaries	30 kg	30 kg	
battery weight	114 kg	190 kg	- 40%
Total weight	804 kg	923 kg	- 13%

However, the check of the inequality (2) demonstrates that the installed battery capacity is insufficient in this instance to suitably accelerate the vehicle with this type of battery:

$$Mb = 114 < (804 - 250 + 70)\ 20/90 = 139$$

Under these conditions, it is necessary:

– to increase the weight of the battery by 25 kg to have sufficient power with the same specific power;

– to develop a battery that is not only higher-energy but also more powerful. In this example, it would be necessary for the specific power to be at least equal to 110 Wh/kg.

Conclusion

The three fundamental conditions of an urban electric vehicle which determine the value of the vehicle weights are: payload, acceleration and autonomy.

The payload and acceleration capacity require a minimum battery weight to have the necessary power. The autonomy determines the battery weight likely to provide the energy necessary to achieve this performance.

Future batteries must have not only high specific energy to make it possible to reduce the battery weight needed, but they must also have a high specific power to be able to have the same power with a low battery weight.

3.3. Recharging systems for electric vehicles

Introduction:

The type of charger and the nature of the recharging infrastructure strongly depend on the kind of vehicle and its use. In this chapter we look at the various possible recharging methods and their scope of application, and we will then describe in detail the infrastructure, the charging stations and the difficulties related to their establishment, and lastly we deal with the safety and standardization aspects.

3.3.1. *What is battery charging?*

During vehicle use the battery transforms chemical energy stored in its electrodes by an oxidation/reduction reaction into electric energy. At the time of the recharge the process is reversed. The charger imposes a higher voltage than the battery voltage and an opposite current. The current provided by the charger will be controlled with precision in order to impose charging conditions required by the battery.

This requires a certain number of measurements (voltage, current, temperature) which are interpreted by the electronics of the charger. The simplest systems are only controlled by the battery voltage: the current is kept constant as long as the battery voltage does not exceed an initial threshold.

Beyond this first threshold, the battery voltage is kept constant. After this phase, we start the balancing charge where a lower current is applied. All battery manufacturers recommend a specific charge algorithm according to the battery service operation, which should be respected to optimize the operating life of their product.

The selected charge algorithm will take account of the batteries' service requirements. Indeed, optimum management of the batteries requires a close link between the manufacturer of the battery and the integrator of the vehicle that defines the charger specifications.

3.3.2. *The various types of chargers*

In the field of road vehicles we can distinguish five types of chargers:

The industrial charger, the low- or average-power on-board charger, the external fast charger, the inductive charger, and the charger with automatic connection.

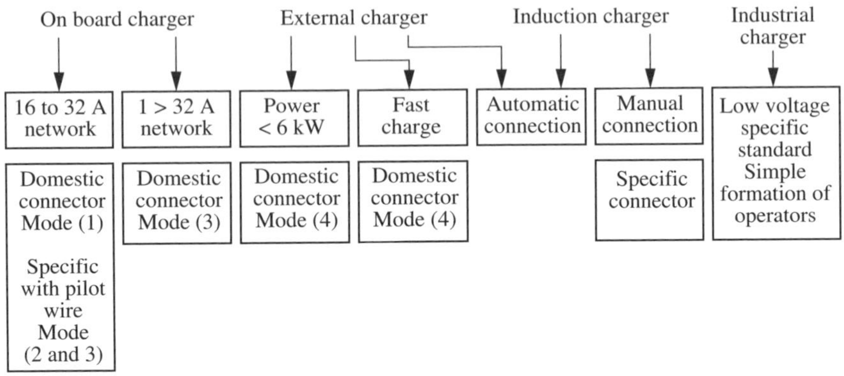

Figure 3.1. *Panorama of charge system*

The external charger supplying the battery directly

This is the traditional charge method used for all industrial traction battery applications. Each pack of industrial vehicle batteries is connected to this type of charger.

Some details concerning recharging in industry

The traditional recharging of traction batteries is done by an external charger connected directly to the battery by cables using grips or low-voltage connectors. It is this method which was adopted for the first electric vehicles at the beginning of the 20th century, and remains today the method preferred for industrial trucks and other industrial vehicles.

When the number of electric vehicles exceeded the number of thermal cars, recharging was done directly via external chargers installed in garages.

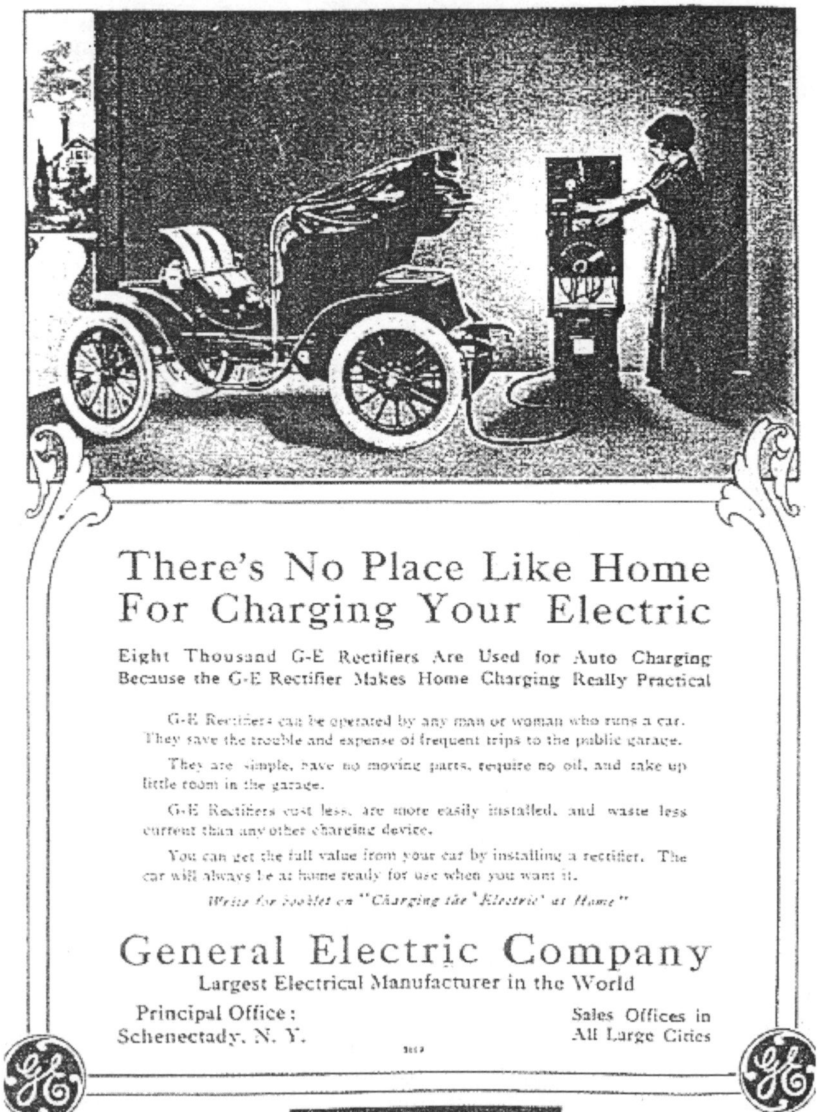

Figure 3.2. *General Electric advertisement for a very simple charge system*

The recharging of traction batteries for the majority of industrial applications is performed by an external charger connected to the electric network. The connection between the charge equipment and batteries is carried out by two flexible wires with a two-pin specific connector.

In addition to the two high-power contacts, the connector can comprise a link for protection wires ground link and for the realization of additional controls with auxiliary contacts.

It is the method which is most generally adopted for the recharging of utility vehicle fleets and small golf cars. This system, rustic but robust, was useful for the first experiments on electric vehicles whose technology was directly based on industrial vehicles.

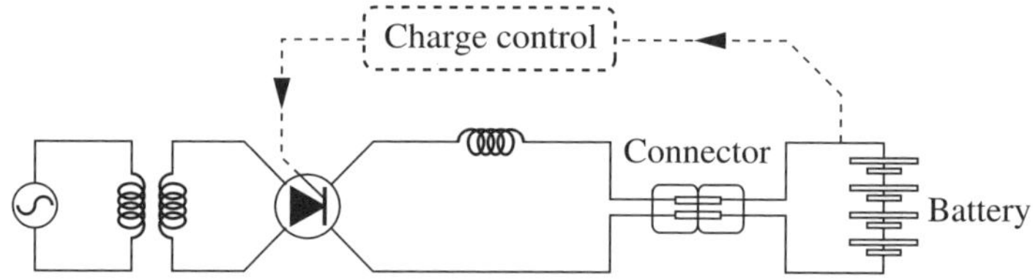

Figure 3.3. *Diagram of a traditional charger using a 50 Hz transformer*

The on-board charger

For urban electric cars we prefer on-board chargers, which allow recharging on all electrical outlets having a gauge and adequate protection.

The convenience of this solution for the customer largely compensates for the disadvantages related to the on-board additional weight and the increase in cost of each vehicle.

Moreover, the charger is perfectly suited to the battery. The use of external chargers for electric cars will be reserved for fast charging requiring heavy, voluminous equipment.

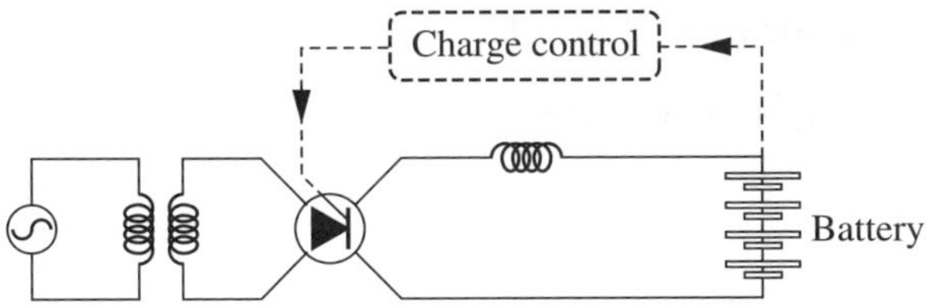

Figure 3.4. *Diagram of an on-board charger using a 50 Hz transformer*

The main constraints of this type of charger:

– volume;

– insulation between the network and the battery;

– the shape of the absorbed current;

– the use of a domestic plug (also dealt with in standardization);

– the maximization of the power available on a domestic wall-plug.

Today, chargers are of the switching–system type (PWM, pulse with modulation).

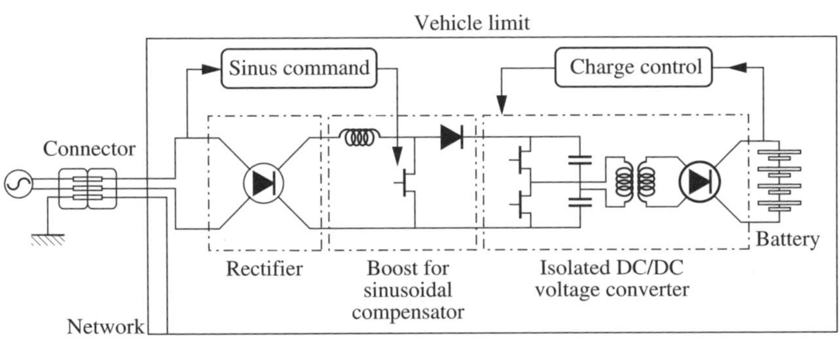

Figure 3.5. *PWM charger*

Make-up of a current electronic charger:

1 conditioning of the network current (rectifier);

2 power oscillator;

3 high-frequency transformer and rectification.

The evolution of chargers in the future will see an increase in power, and/or a reduction in volume and a reduction in cost.

If we accept a non-insulated charger, we can consider the use of the motor inverter and motor coil to carry out the charge function.

Fast charging

Fast charging allows 80% of a battery to be recharged in 20 minutes, this requiring an external system and high power (20 kVA and 400 A).

On-board fast charging

Physical limits to fast charging:

– maximum current permitted by the battery;

– limits to the charging (between 50 and 80%);

– limits according to the type of battery;

– maximum current permitted by connectors.

Type of circuit used for the charger:

(50 Hz and high-frequency transformer, three-phase current, compensation of the power-factor). The need for a particular dialog protocol between the vehicle and the charging station.

Type of connector:

Need for a bulkier specific connector (subject covered in the standardization section).

Induction charger

Recharging by induction makes it possible to transfer energy to the vehicle without requiring physical contact between the vehicle and the charger. Energy is transmitted by a magnetic field through a transformer with separable primary coil and secondary coil. Although it is possible to carry out an energy exchange with transformers operating at the frequency of the electric network (50 Hz or 60 Hz) it is more common to use a high-frequency oscillating circuit to provide this function in order to reduce the volume of the transformer coils. Indeed, the power that can be transmitted through a transformer air-gap is proportional to the induction (field intensity), the transfer and the frequency. It is thus useful to increase the operating frequencies of the systems and to maximize the power of the field in order to reduce the volume of the coils.

The operating principle of chargers with high-frequency induction is similar to that already indicated for on-board electronic chargers and the electric diagram can appear more or less identical.

The diagram is divided into two:

– the network rectifier, the power-factor corrector and the high-frequency oscillator are housed in the equipment on the ground;

– the connection between the vehicles and the charger is carried out by the transformer without Ohmic contact;

– the high-frequency rectifier and the charge detection system are housed in the vehicle.

It is noted that this diagram requires an information exchange protocol, as is the case for the fast charger.

Note: the diagram of 50 Hz supply with the rectifier is given as an example. This functionality can be carried out in various ways (three-phase current supply, passive filtering, etc.). We can use other types of high-frequency oscillators.

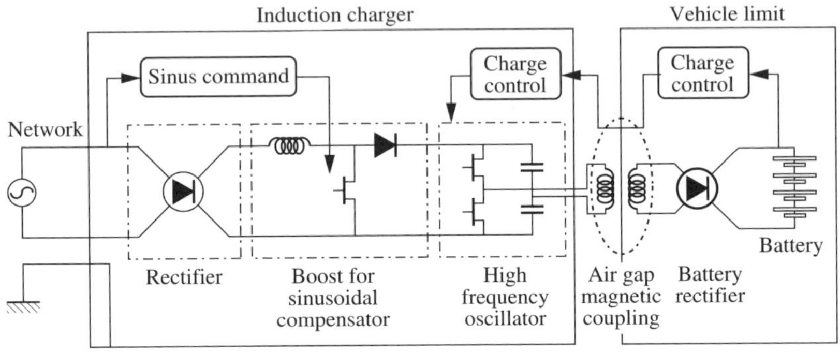

Figure 3.6. *Diagram of an induction charge system*

Application of induction to replace the plug (manual plug-in)

Advantages and disadvantages brought by induction for manual connection:

The efficiency of the recharging by induction lies between 40% and 70% according to the circuits and the distance between two windings.

The dimensioning of the elements is more critical. The transformer is the most complex element and requires the use of ferrite to channel the magnetic flux.

The difficulty is in producing transformers with a good coupling which admit a high air-gap.

Application of inductive recharging for automated connection

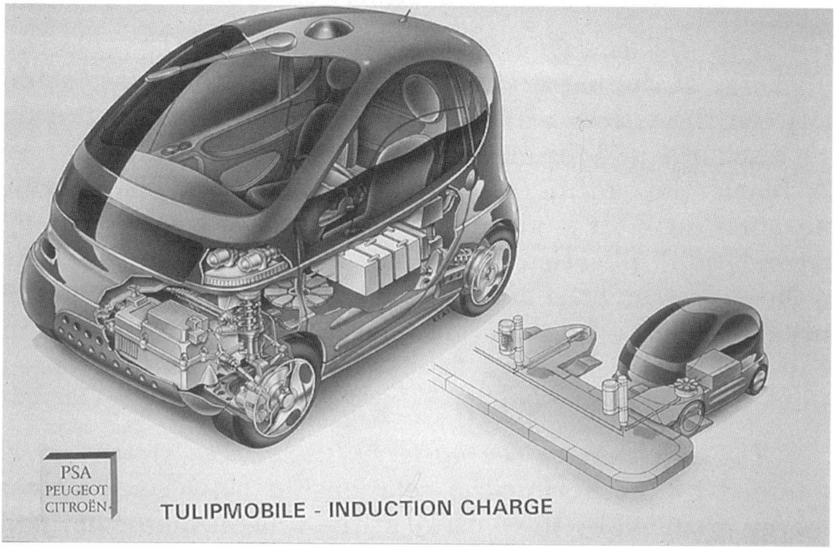

Figure 3.7. *Induction charge system; Tulip prototypes from PSA Peugeot Citroën*

3.3.3. *Recharging efficiency*

Like all physical transformations, charge and discharge do not have an efficiency of 100% because the chemical processes are not perfectly reversible and the battery has certain internal impedance which dissipates part of the energy in heat. We retain three efficiency coefficients:

– *The faradic efficiency*, which is equal to the relationship between the number of coulombs (the integral of the current) provided by the battery during the discharge and the number of coulombs necessary at the time of the recharge to reach the initial charge state of the battery before the beginning of the discharge. It is the coefficient generally indicated by manufacturers. This is close to 1 for new batteries (nickel-metal hydrire, lithium ion). It is a little lower for lead-acid batteries and open nickel-cadmium batteries, which require a longer overload phase.

– *The electric efficiency* of the battery. The electric efficiency of the battery, which is given by the relationship between the energy actually provided by the battery during the discharge and the electric energy provided by the charger to reach the initial state of the battery. This energy, measured in Watt is given by the integral of the "voltage by current" multiplication at the battery terminals at the time of the charge and the discharge. This value is lower than the faradic battery efficiency, and can pass from 70% for certain lead-acid batteries to 95% for new-technology batteries used under optimal conditions.

– *The efficiency of the battery-plus-charger system.* It is obtained by measuring the relationship between the total energy provided by the battery at the time of the recharging and the electric network energy provided to the charger to reach the initial charge state of the battery. This coefficient thus takes account of all off the losses of the "battery + charger" system. The efficiency of the charger is very dependent on the technology employed and their charge operating process. It is thus a culmination of all the efficiencies.

3.3.4. *Recharging in complete safety*

Standardization

The electric car is at the crossroads of two worlds. When it moves it is a car which must be conceived to respect all the regulations of the automotive world. When it is being recharged, it is also comparable to electric equipment and must respect the corresponding regulations and standards. But it is not equipment like domestic devices because it is subjected to a harsh environment and must resist stronger constraints than those met in domestic usage. The questions relating to electric safety during recharging were largely discussed during the 1990s and gave rise to a series of standards describing charge equipment specifically. These standards are the result of an international dialog which succeeded in establishing a common text which covers the various types of networks throughout the world and which adapts to the various practices of each country. Indeed, not all electrical communications resemble each other and the methods of protection available are not identical.

The standards distinguish various recharging modes

Four modes of direct recharging (using connectors) are defined. The recharging must be done at home, in the workshops of companies and in the street. As an example, in the USA the domestic voltage is only 110 volts but electric vehicles have higher performance and require a higher on-board energy (the electric *Ranger* has 30 kWh of on-board battery) than small French sedans which comprise a battery of only 11 kWh and are supplied from a network having a voltage of 230 V.

Hydrogen emissions

The regulation asks for an absence of hydrogen release under conditions of normal charge.

Various plugs used for recharging

The standardized 16- and 32-amp home plugs

The standard power-plug, reference 309

The plug on the vehicle must allow normal charge on a 16 A plug and fast charge on specific terminals. Integration on the vehicle is an essential aspect. Specific plugs have been developed and standardized.

Figure 3.8. *Maréchal socket on electric vehicles*

Charge stations in public spaces

For example, many chargers are available in France such as: underground chargers, recharging at the place of residence, and calibration of the plugs

Difficulties relating to the installation of chargers in public spaces:

– analysis of the recharging cost;

– cost of installation;

– cost of invoicing;

– difficulties caused by the invoicing.

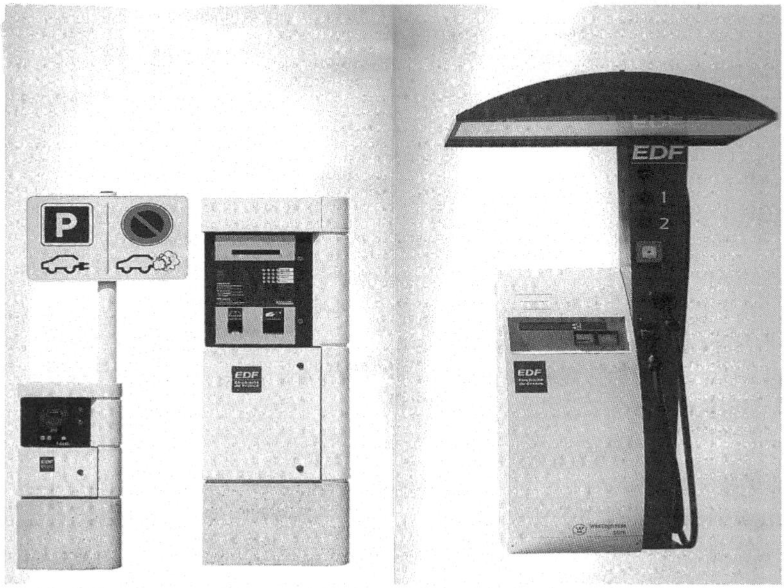

Figure 3.9. *Charge stations*

3.4. Thermal/electric hybrid vehicles

3.4.1. *Assessment of traditional motorizations*

To draw up an assessment of the traditional motorizations is not without importance because they represent at the same time a competitor and a supplier of technologies for thermal engines used in hybrid architectures. It is thus interesting to highlight the latest developments, both in engines and in traditional power transmissions. In addition, we will endeavor to reveal the points on which hybridization will be able to have an appreciable effect.

3.4.1.1. *Recent evolutions and prospects*

Concerning passenger vehicles, the concerns orienting research in recent years mainly relate to the following points:

– reduction of fuel consumption, and consequently of CO_2 emissions;

– improvement of drivability and comfort requirements;

– monitoring of atmospheric pollution and noise emissions;

– reduction of the vehicles manufacturing costs.

While the order given to these priorities can vary according to the situation, fuel saving has very often appeared as the main motivation highlighted by manufacturers and suppliers over the last few years. A parallel can be drawn here with the problems of the hybrid vehicle as is presented in the context of very strong economic competition. Manufacturers will therefore try as far as possible to cut the additional cost induced by technologies implemented to respond to the external constraints which are imposed on them (sustainable development, pollution) by an increase in value brought to the services offered to the vehicle user (drivability, performance, safety).

In the field of engine pollution control, considerable advances have already been carried out with optimized traditional technologies. Concerning gasoline engines, as an example, back in 1997 Honda marketed a version of Accord in Japan given as reaching level (2.3l VTEC engine with 32 bit ECU) whereas Mercedes announced in 1998 a vehicle reaching SULEV level (ML320 with 3.2l V6 engine). Research necessary to these achievements was, however, carried out with an objective of maximum performance on regulated emissions, without any particular consumption constraints (context of before Kyoto). The situation is quite different at the present time where the tendency is an important reduction of consumption while lowering emission levels further with the new European standards planned for future years (see Appendix 1). The challenges to be taken up by motorists are thus very important, especially if it is considered that, in many cases, the improvements obtained on one component can degrade others.

Without claiming to be exhaustive, we indicate hereafter the main axes of recent evolutions or those in progress which we can imagine will have an impact on engine performance and future architectures.

It may be noted that the various points discussed in the continuation can be cumulated on the same production, leading to more significant gains for the same vehicle.

3.4.1.1.1. Gasoline engine

Direct injection

For these engines, one of the main innovations relates to the development of direct fuel injection, where the fuel is directly injected in the cylinder, which was initiated by Mitsubishi in 1998, followed by Renault, PSA, and VW. In contrast to indirect injection (injection into the intake manifold), which is currently generalized, the advantages brought by this technology are mainly:

– A more precise and more flexible control of fuel metering in the combustion chamber brought by the control of the initial time, the duration, the pressure or even the number of injections. The use of a much higher, adjustable pressure (30 to 100 bars for PSA and even 120 bars for Daimler-Benz against just a few bars for indirect) also makes it possible to facilitate the vaporization of the fuel.

– A limitation of the pumping due to the rolling caused by the air valve, mainly with partial loads. The engine can indeed draw a large volume of air (and/or recycled gases – see below), the load being controlled by the quantity of fuel injected. Gains in specific consumption able to reach 20 to 40% at low load were measured on an engine presenting a large operating zone with the air-valve fully open.

– A better dispersion of the droplets, without dampening of the walls, allowing a reduction in consumption and HC emissions.

– A better control of the rattling, making it possible to work at a higher compression ratio with a better thermodynamic efficiency (11.4 for PSA, 11.5 for Audi, 12.5 for Mitsubishi).

– Greater improvements to performance in low-speed torque operation (flexibility) and in the engine dynamics (very low injection reaction time).

The capacities for overall reduction in consumption of such a system are about 10 to 15%, with also a notable increase in the flexibility and power delivered by the engines.

Lean burn

The catalytic post-treatment of exhaust fumes, made necessary by compliance with the emission standards, rests on the principle of obtaining simultaneous oxidation of CO and HC as well as a reduction of NOx on the catalyst system. Such a constraint imposes precise proportioning of the air-fuel mixture, making it possible to construct all of the reactions in the post-treatment catalyst (14.7 g of air for 1 g of gasoline – known as stochiometric richness = 1). It would seem that for uses with partial load and low-engine speed, a lower air-gasoline ratio (richness lower than 1, known as a lean burn mixture) can be used, leading to a reduction in consumption. The triggering of combustion will require, in this case, a mixture of richness 1 within the immediate vicinity of the ignition point, which is generally obtained by a geometrical configuration of the injector, piston and chamber (laminated load). The main issue in the lean burn mixture operation is the very great difficulty in reducing NOx in exhaust gases then rich in oxygen: projections were carried out in recent years in this field with the development of NOx trap systems. The principle consists of storing NOx by adsorption on a complex salt (barium, platinum, palladium and rhodium) then reducing them during richness-1 engine operation sequences. The exhaust line will thus include a precatalysor for the treatment of CO and HC as well as a second element ensuring the storage-release of NOx. We can make the following remarks:

– the reduction of NOx during the release requires operating phases with richness 1, (a few seconds every minute) which will reduce the gain in consumption;

– the operation of the engine with high loads imposes an operation with richness 1, control must return to homogenous mode to the speed generally above 3,000 to 3,500 rpm;

– the NOx trap system is very sensitive to sulfur present in the fuel because the latter is adsorbed preferentially, thus limiting the possibilities of NOx storage, the desorption of SOx created also requiring passages with richness 1;

– the storage capacities of NOx are limited at low and very high temperature;

– the control of the desorption phases requires a precise evaluation of the stored quantities, therefore the presence of sensors and an ECU (reliability over time, self-diagnosis, etc.).

The total NOx treatment output by these systems is about 75%. In addition, solutions of continuous treatment are also being studied, but with lower outputs, about 40% on average (Mitsubishi).

Prospects for reducing consumption from 15 to 30% for direct injections with lean burn mixture have been announced, however, it can be noted that the gains in consumption will be very dependent on the vehicle usage conditions.

Indeed, nervous driving, with high engine loads, can appreciably reduce the gain.

In fact, direct injection gasoline currently seems to be in a position to conquer a significant share of the market.

It will be noted that the use in parallel hybrid, coupled with an electric motor, enabling high-load zones to be avoided, should increase the potential of this solution.

Achievements

These technologies have been implemented over the last few years by manufacturers, alone or close couple as the following examples illustrate:

– the Renault F5R 2l IDE engine which implements a direct injection operating in stoechiometric mixture obtained by recirculation of the exhaust gases (EGR). The interest of this solution is in being able to preserve the traditional three-way catalyst pollution control system. The consumption gain is said to be about 15% on a European standardized cycle;

– the PSA 2l Hpi engine which operates in direct injection and lean burn mixture with NOx storage-release catalyst system. This engine is said to allow a consumption gain of about 10% (including the phases of NOx and sulfate release) with compatible Euro 4 (2005) emissions levels;

– 1l VTEC engine from Honda which equips the Insight and operates in indirect injection with lean burn mixture. This engine, which uses a specific NOx storage-release catalyst, is said to be at the origin of 30% of the vehicle's consumption gain, which is announced by Honda with 3 l/100 on a Japanese 10–15 mode cycle;

– the Audi FSI 2 and 1.6l engines which operate with lean burn mixture and combine a gain in power (15%) and consumption (15% announced) compared to indirect injections.

3.4.1.1.2. Diesel engine

In the field of automobile diesel engines, the innovations relate primarily to the control of injection and emissions (particles and NOx).

We can mention:

– for injection, the marketing of direct injection and the use of common rail injection have made it possible to control the pressure-time-flow parameters, leading consumption gains of 10 to 20%, with an improvement in drivability compared to the previous generation. A second generation of injection systems is now proposed with the following objectives:

- a better control of the injectors (valve close to the opening or piezoelectric control) and of their dynamics (reduction of switching times);

- higher, modulated pressures (1,600, 1,800 and even 2,000 bars announced);

- a greater number of injection holes (six or even nine announced) with a reduction of their diameter (140 to 100 and even 50 μm), in particular allowing a better diffusion of the jet;

- a greater number of injections (up to five) with the possibility of differentiating main injection, pre-injection and post–injection;

- a management of the quantities to be injected by a high-performance ECU (32 bits, 40 MHz).

These evolutions will make it possible to limit noise and NOx emissions, combined in particular with the use of a cooled EGR.

For post treatments, the oxidation catalysts already used could be reinforced by using DeNOx catalysts and particulate filters, the latter making it possible to reduce by a factor of more than 100 the quantity of particulates emitted in an equivalent diameter range of 10 to 500 nm. Such devices will make it possible to comply with the future Euro 5 & 6 standards.

3.4.1.1.3. Downsizing

This process consists of using an engine with reduced cubic capacity, strongly turbocharged in order to preserve significant air and fuel quantities. In order to be sufficiently powerful the engine must use a very dynamic turbocharger-making it possible to limit the response (turbo lag) at least; a turbo with variable geometry can be used or even, in the future, an electrically-assisted turbo. Applied to a diesel engine this process has made it possible, for example, to replace a 1.9l atmospheric engine developing 51 kW with a 1.4l turbocharged engine developing 55 kW.

This evolution is very well adapted to parallel-type hybrid use for the following reasons:

– the torque of the electric motor, available at very low speed, makes up for the reduction in torque of the thermal engine, the consequence of a lower cubic capacity. In addition, the extremely fast rise in the electric motor torque makes up for the turbocharger response-time. This association thus makes it possible to retrieve the flexibility and the dynamic performances of the original engine;

– the reduction of the cubic capacity makes it possible to decrease the volume and weight of the thermal engine, thus facilitating the integration into the vehicle of the electric motor with its control and its power battery (see section 3.4.2 on implementation).

3.4.1.1.4. Auto-ignition combustion engine

This novel mode of combustion can be derived from combustion by compression (diesel); it is a case then of the spontaneous combustion of ultra-homogenous lean burn mixture (Homogenous Charge Compression Ignition – HCCI) or combustion by positive ignition, where it is a case of a controlled spontaneous combustion (Controlled Auto Ignition – CAI).

These modes of combustion are obtained by an injection very early in the cycle for the HCCI or the combined effect of the heat from the burning gases and the compression ratio for CAI. As for the case of the lean burn mixture, these particular operations are not easily applicable to uses with strong torque and strong power; they could consequently be adapted for use in a parallel-type hybrid architecture.

This type of operation requires, in particular for CAI, a variable control of the valves which may be provided by an electromagnetic control. The consumption performance is very good, with maximum gains on the bench from 20 to 30% and prospects for use from 10 to 15%. In addition, the fall in the combustion temperature, due to the creation of a multitude of hearths, leads to an NOx emission reduction at the source of more than 90%, as well as a very significant reduction of the particulate emissions for the HCCI.

The diffusion of this type of engine could take place around 2009–2012. There would be a convergence of gasoline and diesel technologies with the possible use of a single fuel.

As for the lean burn mixture, these new processes will adapt well to hybridization, whether for the limitation of strong loads or high electricity consumption (solenoid valves, etc.).

3.4.1.1.5. Electromagnetic control of the valves

There exists a significant potential for gain on the thermodynamic engine cycle by more precisely controlling input and output gas flows; this can be obtained by controlling the time, the displacement duration and the stroke of each valve. Manufacturers have already developed systems allowing more flexibility in valve movement while keeping the principle of camshaft control (Valvetronic from BMW, i-VTEC from Honda, VVT-i from Toyota, etc.). The electromagnetic control makes it possible to open or close each valve independently, this at any time during the cycle. In the majority of productions the movement of the valves is ensured by an actuator including a mass-spring system and one or two coils controlled while running. Such a system, which no longer requires a camshaft, makes it possible to provide the following functions:

– practically completely eliminate the use of an air valve for controlling the load of a gasoline engine, thus increasing the engine-cycle efficiency;

– allow a better control in the use of multi valves;

– allow a better control of the internal recirculation of gases in the cylinder;

– lower the idle mode and thus the corresponding consumption (about 20%);

– disconnect cylinders of an engine which we do not want to use during low loads (engine with variable cubic capacity);

– increase the torque available at low speed;

– limit the number of engine cycles at low loads; tests have thus been carried out with a 12-times cycle, that is, 4 traditional times plus 4 turns of the engine without combustion and with closed valves. Consumption gains on the bench of more than 20% were witnessed under these conditions;

– vary the compression ratio, for a CAI-type operation or one with prolonged expansion (Miller-Atkinson cycle used for Toyota Prius).

A combined use of these functions is possible in order to optimize the operating range as a whole; the total consumption reduction potential is from 10 to 15%, a value of 15% having been demonstrated by Renault for a 2l 16-valve engine on a European standardized cycle.

While the advantage of camless engines is becoming apparent, their diffusion is still limited by a significant number of problems not yet solved; we can list in particular:

– high electricity consumption, about 1 kW for 8 valves to 2 kW for 16 valves;

– the control and synchronization, which requires the use of powerful ECUs (32 bits) and fast networks to ensure operation during failure modes;

– the significant acoustic emission due to high valve squeezing speeds;

– high volume and weight of head cylinders;

– reliability and cost problems related to technological innovations.

Taking into account their significant electricity consumption and their high engine-control potential, these systems are very well adapted to hybridization.

3.4.1.1.6. Transmissions

The limited revolution speed range of thermal engines requires the use of a gearbox. Various systems currently coexist, the simplest being the manual gearbox (MGB), which also exists in a robotized version with a control of the ratio and clutch changes (RGB); finally the gearbox can be automatic (AGB) or continuously variable (CVT).

The current trends are leading manufacturers to seek stronger integration between engine and gearbox as well as automatic control of the gearbox, making it possible to better use the thermal engine while improving car drivability.

ZF thus proposes a 6-ratio AGB, which would allow a reduction in consumption of 5 to 7% and FIAT proposes a highly integrated engine-RGB unit allowing a reduction in consumption of 10% in mixed cycle. Work is also in hand with an aim of reducing torque lags during gear change in order to improve car drivability. Suppliers such as Luk and Borg Warner thus propose GBs using two clutches and two input shafts, for the even and odd ratios (this gearbox can also integrate an electric motor).

The CVT also represents an interesting pathway for the total optimization of a drive train with a potential of reduction of about 10 to 15%. It seems, however, that engineering problems (volume, cost, lubricant, maximum torque) have limited its use to a few marketed models.

With respect to hybridization, it appears that synergies could be obtained by integrating an electric motor into the transmission and by its use – in addition to the propulsion – to ensure additional functions like synchronization, the supply of torque during gear changes or the drive of auxiliaries during stop phases.

3.4.1.1.7. Electrical equipment

The number of auxiliaries and electric motors on-board vehicles is continually increasing, to a significant extent in the last few years, and a current top-of-the-range vehicle can thus have up to 120 auxiliary electric motors. The on-board electrical network is thus requested more and more for equipment relating to passenger comfort, drive-train control, safety or communications. Consequently, the electric power of vehicle electrical networks has witnessed an average increase in recent years of about 120 W/year. Amongst the new systems which could appear in the next few years, we can mention:

– the thermal conditioning of the cockpit, which includes air-conditioning but also the heating, the latter at times being necessary in order to reduce the temperature setting time in winter start conditions (the thermal contribution of the internal combustion engine not being sufficient because of its efficiency level or due to the partially/totally electric operation of the vehicle). Such a system was notably set up by Ricardo on its Astra prototype, with 135 kW electric heating and a 3.3 kW electric air-conditioner compressor;

– exhaust gas cleaning by plasma; this method allows conversion rates from 50 to 80% on gaseous pollutants and particulates but requires significant power (possible applications for diesel around 2010);

– electromagnetic electric-valves control with a power from 1 to 2 kW;

– the electrically-assisted turbocharger, a method that makes it possible to appreciably reduce the response time (e.g. from 3.5 to 1 second to reach the maximum speed) but requires a motor being able to turn at more than 200,000 rpm and to operate in the vicinity of gases at more than 1,000°C. We will note the possibility of also using this system on a generator, which further still opens new pathways in term of energy management;

– electric heating of the catalyst or filtration system; this solution is an alternative to the use of a fuel burner, the potential emissions reduction on a gasoline engine in cold start is from 30 to 70%. Powers being able to reach 6 kW, for a particulate filter heating system, are possible.

Such an evolution, towards increasingly powerful on-board power network systems, requiring more voltage, goes completely in the direction of the participation of electricity in vehicle propulsion such as it is proposed in hybrid thermal-electric architectures.

3.4.1.2. *Conditions of use*

In parallel to the technological changes, it appears that the conditions of use of the vehicle and its transmission are a factor of utmost importance with regard to consumption and pollution emissions. The description of the interest of hybridization passes by a systematic study of the real conditions of use of traditional motorizations; we will distinguish mainly the points set out hereafter:

– *Idle:* when the vehicle is stationary, the engine is in use yet it does not produce energy for supplying propulsion (we will see that it can, however, ensure the drive of auxiliaries like the air-conditioner). These phases of use can be non-negligible; studies of real usage carried out at INRETS indicate that downtimes can count for 15 to 30% of the total time in urban conditions and 10 to 20% in road conditions.

– *Deceleration:* in these operation phases, the engine is driven by the transmission; in the past, with engines using a carburetor, consumption was not null during these phases. Now the generalized use of controlled injection allows a more complex management of this operating process with possible cutoff of the fuel injection at high engine speeds.

– *Temperature setting:* the incidence of the start in completely or partially cold conditions of the vehicles (engine and catalyst) on their consumption and emissions is very significant. Indeed, a study carried out at European level highlights that half of the journeys made are less than 3 km, that a third are begun with an engine temperature of less than 30°C, and that a third are made entirely in thermal transient conditions. In the case of the European cycle with cold start, the over consumption observed is of about 15%;

– *Transmission:* the low revolution speed range of thermal engines imposes the use of a gearbox and a clutch. This has as a consequence a degradation of the energetic efficiency (especially on the first ratios) and the appearance of speed and torque transients during gear changes (effect on consumption and emissions).

– *Electricity generation:* in current vehicles, electrical production is ensured by an alternator driven by the thermal engine; consequently, the operating conditions of this alternator are not related to the electric request on the network but to the usual conditions of use of the thermal engine.

The fast increase in the power of on-board network systems, related to an increase in the number of accessories (air-conditioning, comfort, safety, communication) and to the progressive electrification of functions, will further increase the effect on fuel consumption. It is considered that, under the current conditions, the electric production of 100 W involves an increase in emissions of 2 to 7 g of CO_2/km, the average electric output produced on-board a European vehicle being, according to Ricardo, 800 W in 2001.

– *Operating range:* in a conventional vehicle, the thermal engine is dimensioned in order to satisfy the highest needs, that is to say generally overtaking or maximum speed on a slope. However, it appears that the powers put in play during dense urban-, or even peri-urban-type journeys, are much lower, with about 5 to 15 kW for a mid-range vehicle.

Consequently, the ranges of use of thermal engines in urban conditions will be in completely unfavorable zones of their operating range. This is illustrated in Figure 3.10, where we have placed the operation points of a thermal engine in a simulation taking into account various uses via cycles of real use developed at the LTE.

It is noted that when journeys become faster, the range of use increases and moves towards zones of better engine efficiency.

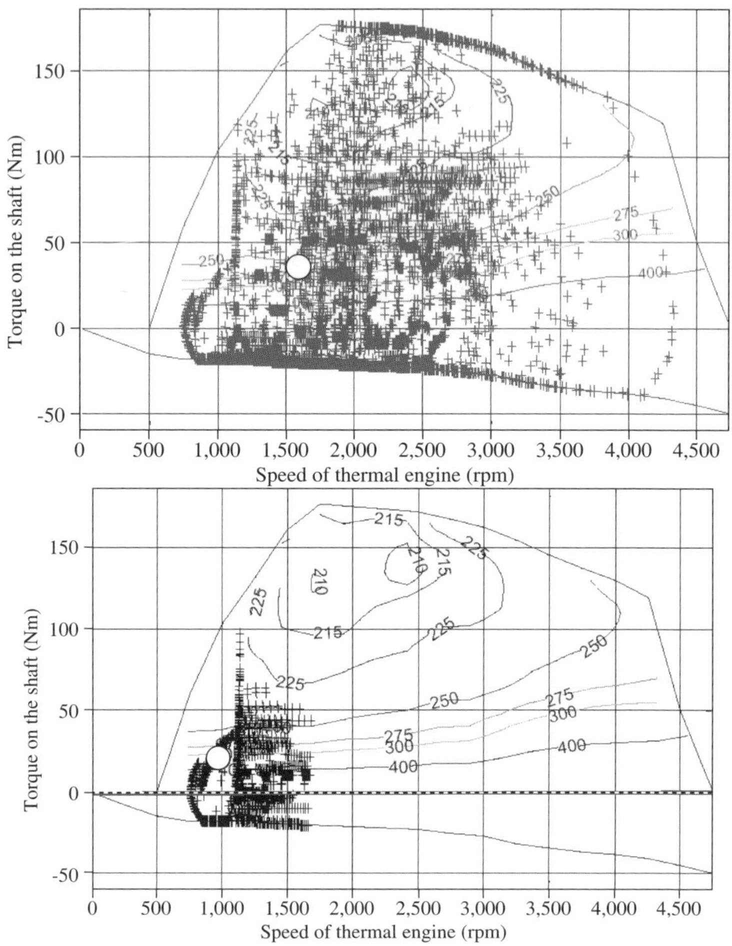

Figure 3.10. *Illustration of the usual conditions of use of a thermal engine for a passenger vehicle, Peri-urban-type use (top) and urban congested (below)*

3.4.2. *Implementation of hybrid transmissions*

3.4.2.1. *General principle*

The motorization of a road vehicle must provide a double function, which is the supply of the power necessary at every moment to obtain the dynamic performance and the utilization of the necessary energy to ensure autonomy. For the conventional transmissions that we mentioned in the previous chapter, these functions are provided by the thermal engine associated with its fuel tank.

In contrast, the principle of hybrid transmission rests on a separation of these two functions, which will then be ensured by different components, specific and optimized, linked to each other and to the wheels by a coupling (see Figure 3.11).

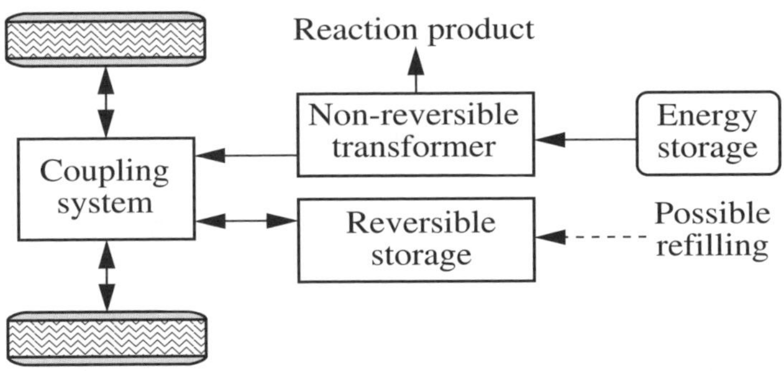

Figure 3.11. *General diagram of a hybrid transmission*

We can thus distinguish the following components:

– an irreversible energy transformer associated with its external energy storage (tank). This transformer constitutes an open system supplied with fuel and combustive and emitting products generated by the internal chemical reactions. It can be a thermal engine with internal or external, alternative or continuous combustion, a fuel-cell generator, or a mechanically rechargeable electrochemical generator such as a zinc-air fuel cell. It will be noted that the transformer will be able to directly energy comparable to that of the storage, in the most favorable case (Table 3.1, cases 1 and 2) or in the opposite case, require a conversion (cases 1, 3 and 5). The solution utilizing two conversions is to be forbidden (case 6);

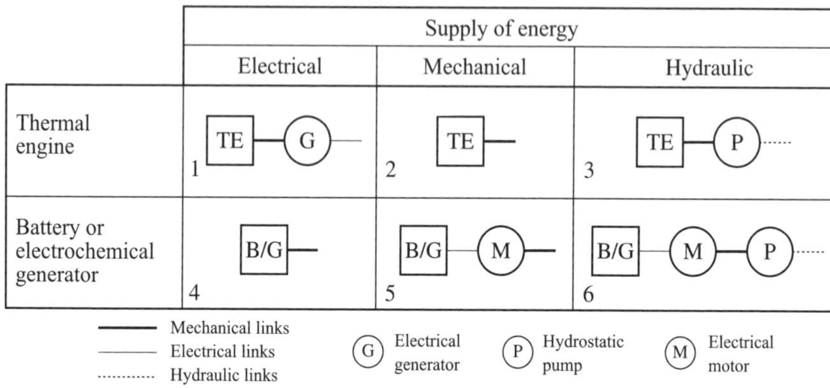

Table 3.1. *Energy transformer systems*

– reversible energy storage. This closed system constitutes a tank which can provide or absorb energy on behalf of the other transmission units. This system may be made up of a battery of super-capacitors, accumulators, of a combination of these elements, or an inertial or hydropneumatic system. This storage will possibly be rechargeable from the outside using the electrical network. As previously, we will distinguish solutions not requiring an energy transformation (Table 3.2, cases A and I), less powerful solutions requiring a conversion (cases B, D, E, F and H) and solutions to be avoided with two conversions;

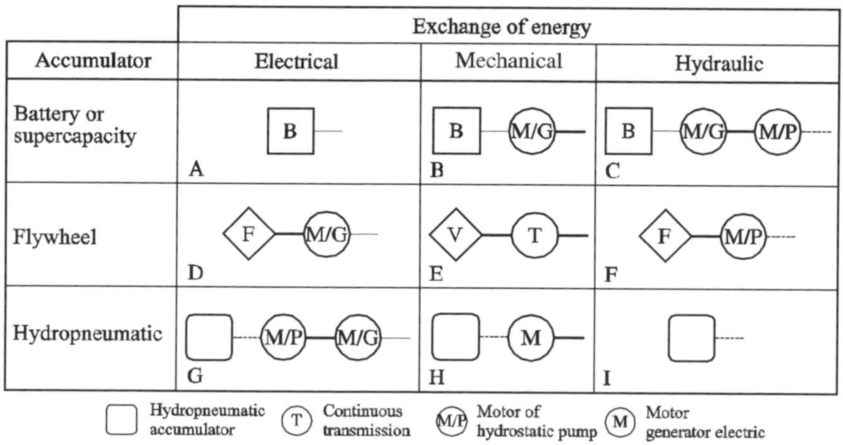

Table 3.2. *Energy storage systems*

– a reversible coupling system. This system ensures the connection between the preceding elements and the drive wheels. The coupling will be able to include elements of a mechanical nature (pulley-belt, chain-pinions, gearbox, epicyclical gear) of an electrical nature (electric machine) or hydraulic (pump). It will be noted that the coupling system utilizes energies of the same nature and must provide mechanical energy to the wheels; it could thus be necessary to convert some of the energies exchanged with the transformer or the storage system (see Table 3. 3).

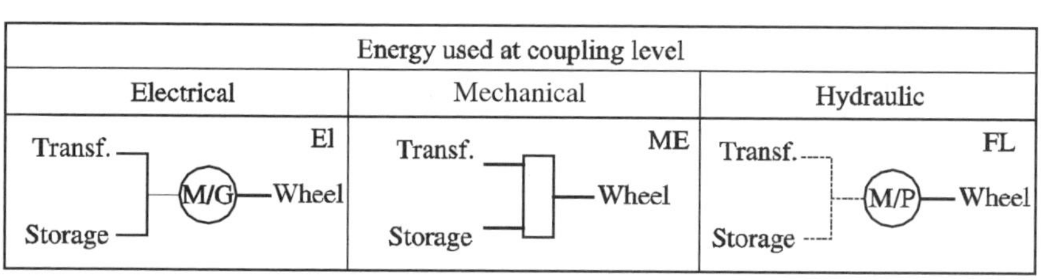

Table 3.3. *Coupling systems*

3.4.2.2. *Simple classification*

A classification of the hybrids can be produced according to the mode of coupling of the two propulsion chains, that is to say:

– series for a coupling putting into play energies of an electrical nature or pressures;

– parallel for a coupling putting in to play energies of a mechanical nature;

– mixed, parallel-series, or dual mode for couplings allowing the two configurations.

A general synoptic diagram of the various possible configurations is presented in Chapter 2. It will be noted that whilst the series-type coupling presents few alternatives – aside from its diesel-electric simplified alternative – the parallel-type and mixed couplings authorize a very large variety of configurations which will be detailed in Chapter 4.

3.4.2.3. *The different configurations*

A great number of configurations were studied in the past for light-vehicle, bus or heavy-truck applications. These projects implemented oil and air storages, flywheels, or gas turbines, in series or parallel configurations, as illustrated in Appendix 2.

It appears, however, that the only projects having led to marketed productions on a large scale in the field of the light vehicles or buses associated a thermal engine as a transformer and batteries as storage. These configurations will thus be considered firstly in the continuation in order for our subject to preserve a concrete character. We will, however, keep in mind that:

– solutions implementing fuel-cell systems are studied for the mid– and long-term, this component being covered in Chapter 4;

– solutions implementing a flywheel or super capacities are also being studied in public transport for buses or cars.

3.4.2.4. *The thermal-electric configuration*

The objectives sought in this association are the following:

– to have two drive trains whose performances are complementary, that is:

- thermal: great autonomy related to the very great specific energy of liquid hydrocarbon fuels (about 10,000 Wh/kg compared to 100 to 200 for an accumulator) and ease of energy recharging (the filling of a tank in 5 mins represents an equivalent power of more than 5 MW),

- electric: a strong torque available from zero speed for startup, a great flexibility in the control, the possibility of recovering part of the energy during vehicle braking, possible operation without emissions of local atmospheric pollutants and with very low noise pollution;

– to carry out an advanced optimization in the system operation, making it possible to combine the advantages of each drive chain while minimizing the impact of their disadvantages. This optimization will rest on the use of powerful ECUs which will endeavor to take into account the use of the vehicle in particular;

– to propose vehicles having additional functions compared to traditional vehicles, in particular in order to support their diffusion, which could be affected by higher costs (purchase and maintenance);

– to allow a transfer of consumption from hydrocarbons towards other primary energies via other energy carriers like electricity or hydrogen in the case of hybrids with a fuel cell and battery.

3.4.3. *Context of research concerning hybrid transmission*

We have indicated that hybrid transmission had various advantages related to low fuel consumption, low pollutant emissions as well as the possibility of transferring energy consumption to other primary energy sources. These various assets were proposed (with more or less weight), over the last few years; we can principally distinguish the phases described hereafter.

3.4.3.1. *The initial epoch*

We find traces of the first productions of hybrid vehicles at the beginning of the 20th century, whilst three competitors: vapor, thermal and electric motorizations, fought to succeed horse-drawn traction for road vehicles. Figure 3.12 represents a series hybrid vehicle, designed in 1902 by *Fisher Omnibus* in California.

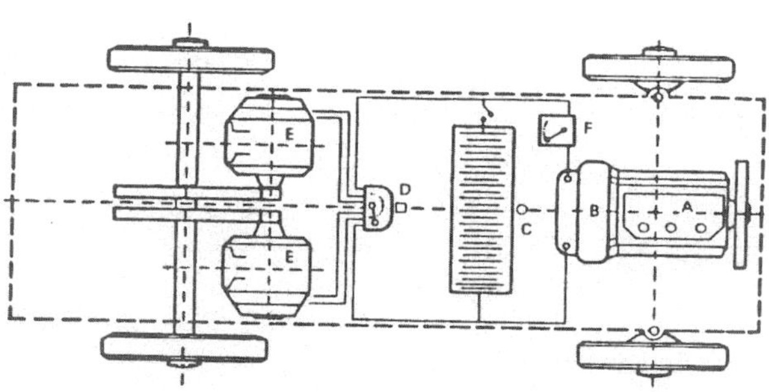

Figure 1. Fischer Omnibus Chassis, ca. 1902.
A=Engine; B=Generator; C=Battery; E=Motor;
D,F=Controllers. (Source: Ref. 1)

Figure 3.12. *Series hybrid vehicle, configuration at beginning of 20th century*

Very fast progress was then made in the field of thermal engines and mechanical drives with, paradoxically, the development of the electric starter, which would make it easier to use the thermal engine, around 1912.

This evolution, combined the with abundance and the low price of oil, gave supremacy to thermal motorization as of the end of the 1910s. The concept of hybrid motorization would thus stay dormant for many years.

3.4.3.2. *Energy management*

The Arab-Israeli War of October 1973 and the ensuing oil embargo brutally emphasized the energy concerns for Western countries. In hindsight, many authors believe that rather than a crisis, these events were a change; the world (primarily the western countries) entered a new era that some people rightly described as the era of limited energy.

With less than 10 years' interval, the Iranian revolution in 1979 and the Iran-Iraq conflict caused what has been called the second oil crisis. Because of these events, the cost of oil increased, in constant dollars, by a factor of 3 in 1973 ($5 to $16/barrel) and a factor of 2 in 1979 ($16 to $32/barrel). Today the 3rd oil crisis is on progress. (More than $120/barrel).

In addition, the very strong concentration of oil resources in the Middle-East alone, which accounts for 66% of world reserves, constituted a fear for the security of the resources of consumer countries; a fear later confirmed by the various wars carried out in Iraq.

Such a change triggered a large amount of research on lower-energy vehicles, that is, by means of concerted R & D programs (vehicle 3l/100 km in France) or standards such as the CAFE (Corporate Average Fuel Economy) set up in the United States and which made it possible to reduce by half the unit vehicle consumption between 1975 and 1989 (often by means of the importation of Japanese vehicles).

This quest for lower consumption vehicles was accompanied by a concern for energy diversification, as appears in the introduction of an article written by an American researcher on the United States research activities at the time in the field of electric and hybrid vehicles, that is to say:

– The oil embargo and the fuel shortages of 1973 and 1974 revived the advantage of the electric vehicle as a means of decreasing the use of oil in the transport sector. Consequently, Congress voted in favor of a subsidy for an R & D program focusing on electric vehicles and hybrids.

– The hybrid vehicle can lead to a 75% reduction in gasoline consumption compared to an equivalent thermal motor vehicle.

– Electric and hybrids vehicles are thus regarded as a means of transferring a share of oil consumption to other primary energies such as nuclear power, coal or hydraulics, using electricity as an energy carrier.

– These concerns can be illustrated by considering two vehicle prototypes developed at that time, which were designed for recharging of their battery on the network, that is to say:

– parallel hybrid prototype developed in 1980 within the framework of a project financed by the DOE. This vehicle made it possible to reduce fuel consumption by more than 50% on the first 60 kilometers, the 13 kWh battery reaching a discharge depth of, 50%;

– Golf hybrid prototype whose first version was launched in 1983 and tested in Zurich on a fleet of 20 vehicles, made it possible to halve fuel consumption, at the price of an additional consumption on the network varying from 9 to 19 kWh/100 km to recharge a 3.6 to 4 kWh battery.

3.4.3.3. *Emission control*

This phase began in the 1980s with in particular a raised awareness of the global effect of air pollution, linked in particular to the mediatization of the phenomenon of acid rain. This new stage could be defined as that of the limited atmosphere.

Parallel to this raised awareness, a certain relaxation appeared on the energy market. This can be attributed to the difficulties of exporting countries (OPEC) in harmonizing their production policy, or to the stabilizing effect presumed due to the presence of Western forces in the Middle-East since the First Gulf War. In addition, the progress achieved in the prospection and exploitation of deposits led to a revalorization of the proven reserves, with the latter indeed increasing by a factor of 1.8 for oil and 3 for gas between 1970 and 1990. The operating-life estimates thus became, at the beginning of the 21st century, more than 40 years for oil, 70 years for gas and more than 200 years for coal (with, however, the assumptions of current consumption).

During these years, vehicle prototypes were developed, in Europe in particular, using a series architecture with a gas turbine (VERT project by PSA and Renault, EEC project by Volvo). Such configurations did not make it possible to appreciably fuel consumption. On the other hand an ULEV emission level, with an autonomy in all-electric mode of 20 to 40 km for urban use was envisaged. Such vehicles had, in particular, the advantage of being able to circulate in urban centers without local pollutant emissions detrimental to residents' health.

3.4.3.4. *Raised awareness of the greenhouse effect*

The greenhouse effect is a natural phenomenon of warming of the low layers of the atmosphere, created by the presence in the air of several gases which have the property to absorb part of the solar energy which is re-emitted by the Earth in the form of infrared rays. This phenomenon is paramount to maintaining life on Earth because it makes it possible to ensure an average surface temperature of about 15°C, whereas the irradiative result alone would expose us to a temperature not exceeding -18°C. The main gases (GHG) are CO_2 (60% of the total activity), CH_4 (15%), CFC (about 12%) and N_2O (5%).

The mechanism of the greenhouse effect has been known since the 19th century, but the first alarmist messages connecting the idea of climate change to greenhouse gas emissions date from end of 1980s. Since then, a great number of studies have made it possible to advance figures which are relatively well admitted by the scientific community, that is to say:

– since the pre-industrial era (1750), the concentrations of the main GHGs in the atmosphere have increased by 30% for CO_2 and 145% for CH_4;

– the earth has heated-up by 0.3 to 0.6°C since the end of the 19th century, at a much greater speed than for what could be reconstituted for the preceding 1,000 years;

– the level is believed to have increased by 10 to 25 cm.

The scenarios carried out, by the work group of the GIEC in particular, report the possibility of the CO_2 concentration doubling in 2100 with consequently a possible rise in the average temperature of 1 to 3.5°C. The consequences would be mainly a rise in the sea level (15 to 95 cm), a change to precipitation and marine currents, with induced damage which could represent up to 2% of the world GDP.

Faced with these threats, in June 1992 in Rio de Janeiro a great number of countries of the United Nations signed a convention aimed at stabilizing GHG emissions. These measurements not being considered to be sufficient, a more voluntary protocol, aiming at a global reduction in CO_2 emissions was adopted in December 1997 at the Kyoto conference. The global reduction ratio of emissions (CO_2, CH_4, N_2O, HFC, PFC and SF6) was fixed at 5.2% for the period 1990 to 2010-2012, the countries concerned being developed countries and developing countries. Distributions were proposed by country, we will note that according to their economic needs, some countries were authorized an increase in their emissions, that is to say:

Japan:	-6%	United States:	-7%	EEC:	-8%
France:	-0%	Germany:	-21%	Italy:	-6.5%
Greece:	+25%	Spain:	+15%	United Kingdom:	-12.5%

It will be noted that large developing countries, currently low GHG emitters, like China or India, are not concerned with these agreements although their potential future emissions are extremely high.

The various energy-consuming sectors consequently developed strategies making it possible to respect these commitments, which proves particularly difficult in a context of economic growth. Concerning road transport, we can make the following observations:

– it is responsible for 20 to 30% of all GHG emissions (26% for France in 2001);

– their share of the total GHG emissions increased significantly in recent years (from 21 to 26% for France between 1990 and 2001);

– evaluations of the future indicate that road-transport GHG emissions will continue to grow, the unit gains on vehicles not compensating for the increase in the distance covered.

Consequently, a voluntary agreement was drawn up in 1999 between the European Community and manufacturers. This agreement concerns European manufacturers (ACEA) as well as the main importers of vehicles from Japan (JAMA) or Korea (KAMA). The text envisages in particular that the Commission will not legislate CO_2 vehicle emissions, because of a voluntary engagement passed with manufacturers, which indicates reduction objectives for the consumption of new cars by the introduction of new technologies. The stages are the following ones:

– marketing of vehicles producing no more than 120 g/km as of 2000 (206 HDi, VW Lupo 3l TDI);

– objective of 140 g/km by 2008 (2009 for JAMA and KAMA);

– objective of 120 g/km by 2012, with evaluation of feasibility in 2003);

– recommendation of 90 g/km by the European Parliament for 2010.

These reductions will represent a fall of 15 to 45% compared to the current values which are about 164 g/km for Europe; we will note that we can already a reduction due to the introduction of new technologies, because the values were close to 190 g/km in 1995 and about 170 g/km in 2000.

In 2008, the objective was not reached, so the EC decided to regulate CO_2 vehicle emissions. Today the regulation has been voted in and it will be in application for 2012 with a level of 130 CO_2 g/km on average across Europe.

It should, however, be stressed that certain major evolutions, like the increase in power of electric auxiliaries, or air-conditioning, are not taken into account in the regulations used to define these values.

The hybridization of motorizations constitutes one of the means of implementing vehicle technologies in the field in order to achieve these goals.

Remarks

As a comparison, the current consumption of vehicles in the United States, which is controlled by standard CAFE, is 9 l/100 km (27.5 mpg or 216 g of CO_2/km) for light vehicles and 12 l/100 km for SUVs and pickups (20.5 mpg or 288 g CO_2/km).

It is appropriate to point out that whilst the reduction in vehicle consumption, and thus in the consumption of fossil products, can constitute an undeniable objective, there are certain highly contentious schools of thought, or certain specific regional geographical characteristics leading to other approaches; we can mention:

– the studies disputing the importance of the greenhouse effect and its consequences, underlining the economic costs of the measures to be taken and placing future questioning on other vital resources such as drinking water;

– the studies recommending focusing on gases other than CO_2 (CO, COV, NOx, CH_4, O_3, CFC) in the fight against the greenhouse effect;

– the case of vast metropoles in Asia or South America or very dense urban centers (Italy in particular) for which local pollution due to transport still induces health issues for residents. On this subject let us note the creation of the Clean Air Initiative for Asian Cities (CAI-Asia), which gathers more than 18 agglomerations and which has the objective of controlling urban pollution. These variations in the nature of the concerns are illustrated in Figure 3.13, which presents the relative importance of the various fields of research in Japan for the coming years.

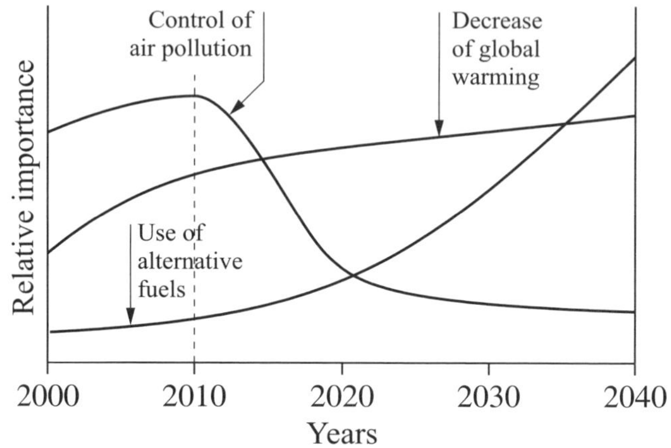

Figure 3.13. *Relative importance envisaged for the various concerns in Japan*

3.4.4. *Functionalities of hybrid architectures*

3.4.4.1. *Choice of classification*

We previously indicated in section 2.2 a simple classification of series, parallel or mixed hybrids. Such a classification no longer appears relevant at the present time, however, in particular for the following reasons:

– the implementation of the mixed hybrid solutions (Toyota Prius in particular) or diesel-electric exceeds simple classification for series and parallel;

– the great number of possible combinations for the only parallel architecture type requires a description in more detail if we want to give an account of all the functionalities of the system.

Moreover, hybrid motorizations are currently in competition with traditional motorizations, whether concerning their consumption, their emissions or their costs (acquisition and maintenance), as such we propose to distinguish them according to the complementary functions which they can bring. It will be noted that these functions will be able to be distributed according two categories, that is to say:

– the functions taking part only in the reduction in consumption and pollution emissions of the vehicle. For the driver, this vehicle will not be distinguished from its thermal-engine counterparts apart from its performance in consumption and emissions; we will then about *discrete hybrids*;

– the functions bringing an additional benefit for the owner (to penetrate into restrictive zones, etc.) people in the vicinity (to limit or cancel local pollution, etc.) or even of the community (to limit the consumption of fossil energy, overall pollution, etc.); we will talk then about *functional hybrids;*

– the various functions brought by hybridization are discussed in the following chapters, from the most basic, implemented in discrete hybrids, to the most complex bringing the originality of functional hybrids.

We will specify for each one of these functions the main applications.

3.4.4.2. *Light hybrids*

3.4.4.2.1. Thermal engine startup (*stop&go*)

Principle:

This function is the most basic; it consists of removing the idle phases, i.e. turning off the thermal engine as soon as it does not have to produce useful energy any more and to very quickly start it up again it in the opposite case.

The function is natural for series hybridization where the starting of the thermal engine is controlled by the energy management system according to the needs.

For parallel hybridization, the thermal engine is therefore stopped when the vehicle stops and even for *foot-off-pedal* decelerations at low speed. In this last case, the engine injection can be cut; it is then driven by the inertia of the vehicle. These evolutions are illustrated in Figure 3.14 for the case of the Honda Insight.

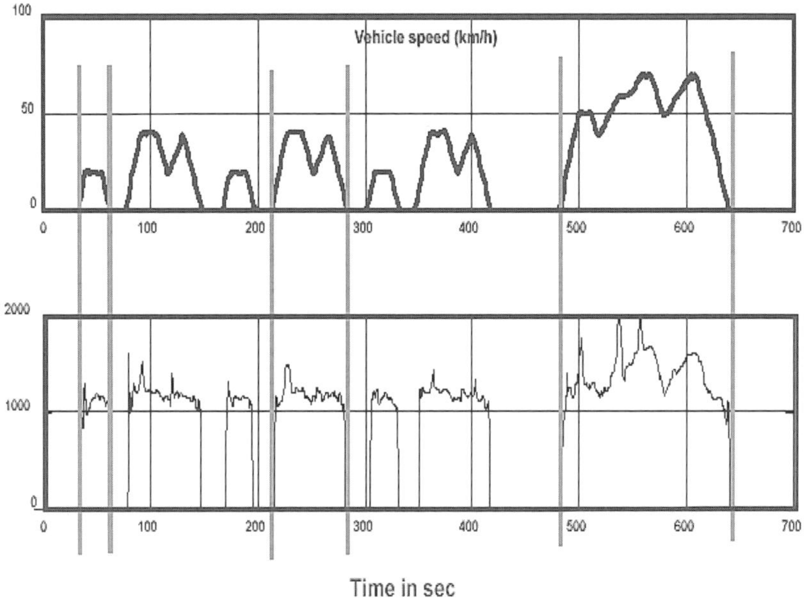

Figure 3.14. *Example of use of the thermal engine of an Insight on a 10-15 mode cycle*

We indicated that the vehicle stop durations used in urban conditions can be significant, consequently the gain can represent from 6 to 15% of consumption in urban use, and will quickly become negligible in road or motorway use. The engine is started as soon as the driver carries out an action on the clutch pedal, the accelerator or on the gearbox selector. The startup time of the system – whatever the conditions on the vehicle and the environment – must be perfectly controlled for obvious safety reasons.

The electric motor in this configuration allows the starting of the thermal engine (*stop&go*) as well as the on-board electricity generation; we talk then about an *alterno-starter*. The following functions must be assured:

– very fast and quiet starting of the thermal engine, which is imposed by rigorous drivability criteria. Startup times, to reach idle mode, lower than 0.5 seconds are sought, compared to 1 to 1.2 seconds for a traditional startup. The very fast acceleration of the engine requires a significant torque to be produced, in particular during cold starts (up to 400 Nm possible);

– on-board electricity generation with high power and improvement of the electric motor and converter efficiency (5 to 10%) compared to the current solutions;

– the operating life of the system: motor, control and battery exceeding several hundreds of thousands startups.

Implementation:

For series hybridization, thermal engine startup and ramp-up can be assured by the generator, which requires reversible control electronics, or by the conventional starter in the opposite case, the constraints on the startup time being rather low.

For parallel hybridization, taking into account the limited number of functions, the electric motor will be low-power and may be linked to the thermal engine by a pulley-belt system, in a non-integrated configuration where the motor takes the place of the alternator (conventional architecture). This solution involves just a minor modification on the propulsion system as we can see in Figure 3.15.

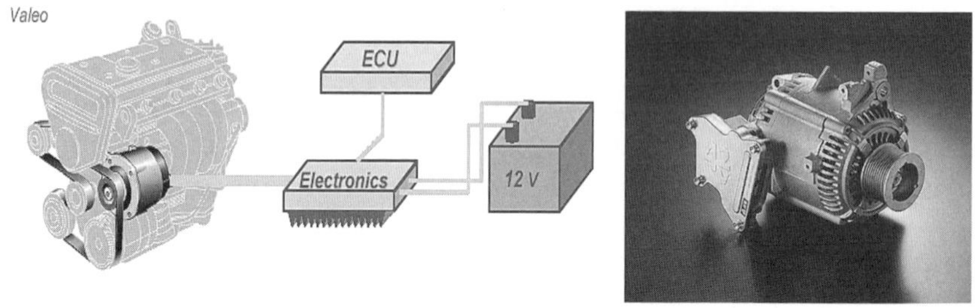

Figure 3.15. *Non-integrated alterno-starter, Valéo document document (Crown)*

Valéo proposes to provide these functions, a 2.5 kW electric motor, cooled by air and connected to 12 V by a traditional starting battery. Such a motor makes it possible to ensure a startup at -25°C for gasoline engines of up to 2.4l, and diesel engines of up to 1.8l. Toyota for its part uses a 3 kW motor connected to 42 V.

The establishment of the starting function requires a system approach taking into count: the thermal-injection-pollution-control engine unit and making it possible to optimize starting in order to avoid the emission of puffs of pollutants (CO and HC). The use of direct injection can be favorable by authorizing a very precise management and better use of the injected fuel. Mitsubishi, for example, on its GDI Sigma *Stop&Go* engine, indicates the first ignition is obtained in 0.12 seconds (1/6 of a turn), thanks to an injection into the first cylinder during it compression phase.

In addition, the suppression of idle phases will induce constraints on the operating of auxiliaries, in particular those necessary to comfort like air-conditioning. The recordings presented in Figure 3.16 illustrate the very limited use of the function as soon as the mechanical-drive air-conditioner is put into action.

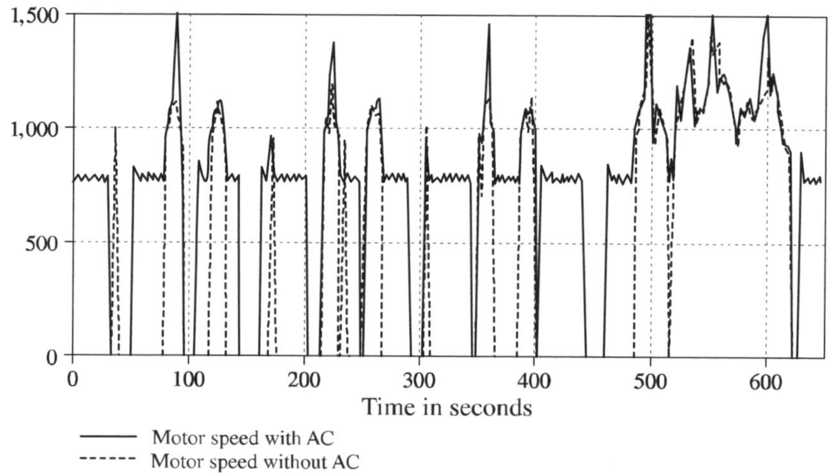

Figure 3.16. *Degradation of stop&go operation with mechanically-driven air-conditioning*

Alternative solutions will have to then be implemented if we want to obtain the maximum benefit from the function without sacrificing the comfort of the passengers; we can mention the following alternatives, which use, during the stop phases:

– a compressor with electric drive supplied by the battery;

– a thermal storage by material with phase shift, solution suggested in particular by Valéo;

– an electric compressor supplied by a fuel-cell generator (APU which can be SOFC and supplied with the fuel of the thermal engine);

– a drive by the thermal engine when the latter is in action and by an electric motor in the opposite case, decoupling being ensured by a clutch (see Toyota Crown in section 3.4.6.1.9).

These solutions present various degrees of maturity; whilst the fuel-cell solution might be for a long time in the future, others such as the electric drive of the compressor are already marketed, as is the case for Toyota on its Crown.

3.4.4.2.2. Thermal-engine assist

Principle:

This function relates more particularly to the parallel hybrid, the series architecture making it possible to partially or completely disconnect the thermal engine from the vehicle dynamic requirements.

Compared to the *stop&go*, the assist function adds the simultaneous use of the thermal engine and electric motor, which will operate in addition to the torque or speed according to the coupling system.

The objective is to benefit from the electric motor action either to optimize the thermal engine (downsizing) or its use, or to obtain an increase in drivability (torque at low speed).

The resulting gain in consumption is about 15 to 30% depending on the configuration and the use of the vehicle.

In complement to the preceding functions, the electric motor can provide assistance here to the thermal engine, generally in torque, during starting (setting in motion or launch assist) or under operation (boost). This simultaneous use makes it possible to implement the following functions:

– use of a thermal engine that is of lower cubic capacity, smaller, lighter, more economical (downsizing) but developing less torque, in particular at low speed;

– better management of the dynamic transients, because of a very low reaction time of the electric motor (suppression of the turbocharger response time, better pollution-control operation);

– increase in the car drivability by obtaining a very strong torque even at low speed (turbodiesel effect), see Figure 3.18; Honda's choice for the Insight;

– possibility of improving automatic gearbox operation by authorizing a faster bridging of the torque converter;

– possibility of recovering part of the energy available during decelerations.

It is important to point out that in this configuration the electric motor and the thermal engine are dependent on each other and it is thus not possible to use one without the other, in traction for an operation in all-electric mode or during deceleration for optimized energy recovering.

Implementation:

The powers put into play for the electric motors in this configuration are about 5 to 10 kW; consequently the two configurations, integrated or not integrated, are possible.

It will be noted that in the integrated configuration, the electric motor's installation in the transmission induces an increase of the overall length of the power-train and thus a constraint in the event of a transverse position in the vehicle.

Detail of electric motor assembled on thermal engine:

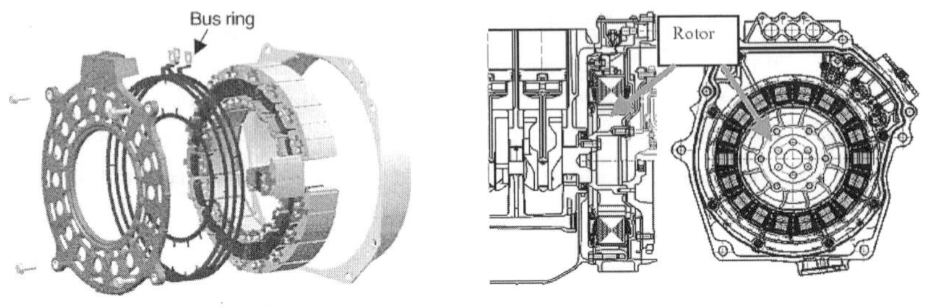

Figure 3.17. *Alterno-starter-booster integrated into Honda Insight*

The thickness of the electric motor is now less than 1 cm/kW with values of about 6 cm to 8 cm for 10 kW, but it appears that a good integration with the clutch can make it possible to limit the total increase length of the engine-transmission group.

In the integrated configuration, the electric motor acts as a starter, flywheel and alternator, which makes it possible to remove these three parts of the power-train.

The booster operation allows various alternatives in its implementation, that is to say:

– lowering the thermal-engine dimensioning and making it more compact, Honda uses a 1 liter three-cylinder engine and Ricardo has developed a 1.2 liter TDI engine (see Figure 3.18). This solution makes it possible to obtain a significant saving in weight, volume and consumption. It should be highlighted that in this case the dynamic performance of the vehicle cannot be preserved if the battery cannot provide any more current (depth of discharge and/or maximum temperature, etc.). According to the achievements, the vehicle will present a relatively significant power loss (12% for the Insight, from 55 to 49 kW) or more limited performance, as on the Ricardo project where the thermal engine was optimized to provide a high torque at high speed, thus limiting the power loss to 8% (80 to 74 kW);

– preserving the initial thermal engine, using it only on its maximum efficiency curve, corresponding to average charges. This operation makes it possible to avoid the unfavorable zones of full load with regard to the pollutants (NOx, smoke), and to consumption (better use of the lean burn mixture). In these conditions the dynamic performance of the vehicle may be preserved regardless of the depth of battery discharge by using, if necessary, the full thermal engine load.

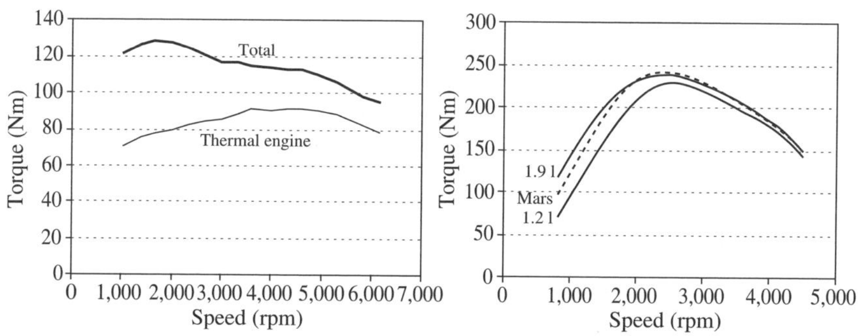

Figure 3.18. *Characteristic of torque with and without Honda Insight electric motor according to Valéo – Ricardo i-MoGen*

It thus appears that for some choices, the dynamic performance of the vehicle will not be able to be preserved; we will then be in a degraded operating mode which can surprise the driver. It is imagined that such a choice is very sensitive for manufacturers it is not decide today, a situation which can be illustrated by the following examples:

– degraded operating modes were highlighted on the Prius in its first version, for very severe conditions of use reproduced on-bench. The version marketed in Europe has since been equipped with a thermal motorization developing more power, which should make it possible to avoid these situations (gain of about 10 kW); the latest version is equipped with a large thermal engine;

– for the same type of pickup, initially by a V8, DaimlerChrysler chose for its Durango a V6 engine associated with an electric motor in order to maximize the gain in consumption, whereas GM, for the same configuration, retained a V8 engine in order to not sacrifice the performance of the original vehicle GMT800 in all configurations.

3.4.4.2.3. On-board generator

As we indicated, the electricity consumption of the on-board vehicle power network does not cease to grow. In this context, the advantages brought by the use of an electric thermal hybrid transmission are the following:

– using high-power electric motors, with an optimized electronic control which makes it possible to obtain efficiency of about 75% and more, significantly higher than those of alternators with a diode rectifier, currently available in the automotive industry;

– having high power high voltage electrical network, well adapted to the rise of the auxiliaries power components.

3.4.4.2.4. All-electric mode

Principle:

In this operating mode the vehicle can be propelled by the electric motor alone, the thermal engine being put into action and stopped automatically by the line supervisor without the possibility of control by the driver.

This operating process is well adapted to series hybridization because the electric motor is dimensioned for the maximum torque and power vehicle capacity, which is not the case for parallel hybridization.

Operation in all-electric mode is illustrated in Figure 3.19 where we have represented thermal engine speed recordings for the parallel and series configurations of the Toyota Prius 1 and Nissan Tino on a Japanese standardized cycle.

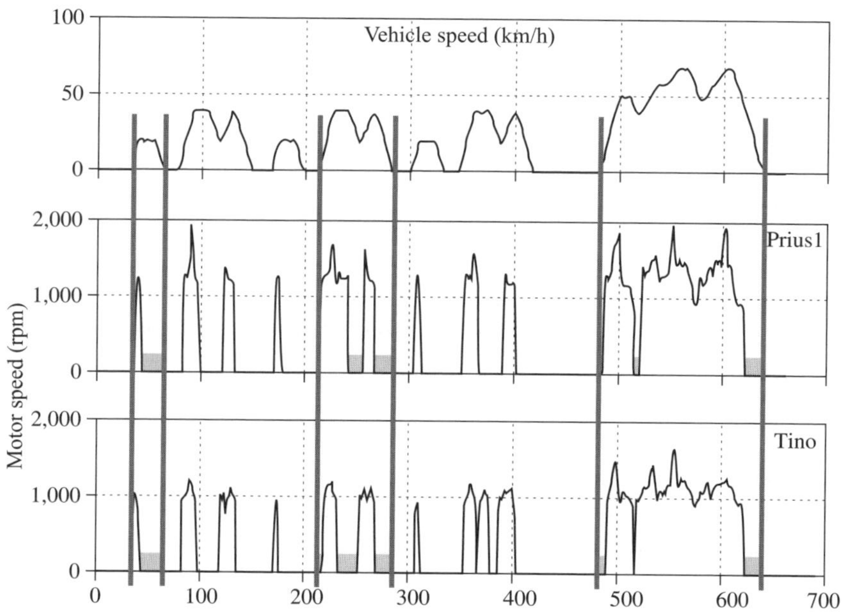

Figure 3.19. *Example of all-electric mode in parallel hybrid (Japanese 10-15 mode cycle)*

It is noted that the evolutions are fairly similar for these two architectures of the parallel type, the thermal engine only being used for accelerations on the sections at low speed, and it does not operate during: the setting in motion of the vehicle, low-speed evolution and decelerations. For high-speed evolutions, we note that the thermal engine is used permanently, except for decelerations.

In the case of series hybridization, the thermal engine can be entirely managed by the ECU and operate with dynamics completely different from that of the vehicle.

Implementation:

This operating mode is natural for a series hybrid.

In parallel hybrid, this use requires architecture with a clutch or a mechanical device making it possible to separate the electric motor from the thermal engine.

We can make the following remarks for the all-electric mode:

– the all-electric mode is managed by the on-board ECU. The parameters used generally include the accelerator pedal position (low demand), the battery state of charge (upper a minimum value), the engine-catalyst temperature (hot) or a request of the auxiliaries (air-conditioning compressor). When one or more of the minimum conditions on these parameters are not respected, the vehicle cannot be used in all-electric mode and the thermal engine will be automatically put into action;

– the vehicle must be able to drive with the thermal engine off, and that implies an electric drive of the vital functions (assistance pumps , etc.). For the comfort functions like air-conditioning, alternative solutions to the mechanical drive will have to be found;

– this operating mode is well adapted to an urban use where the electric motor can give maximum starting torque. In parallel hybrid, we will also have the possibility of ensuring an inertial startup of the thermal engine. In this case, setting in motion in all-electric mode will in addition make it possible to remove the components necessary to the operation of a thermal engine driving the wheels at very low speed (clutch, hydraulic coupler, etc.);

– the constraints on the power battery will be more significant (capacity, cycling, etc.). Energy battery will also be higher (#1.8 kWh for Prius compared to #1 kWh for the Insight);

– recovery during braking is facilitated by the disconnection of the thermal engine.

3.4.4.2.5. Energy recovery during braking

Principle:

The use of an electric motor and a reversible storage of energy in hybrid transmission makes it possible to ensure a recovery of energy during braking or in descents. As for the all-electric mode, series architecture will be favored because of the strong power of the electric traction motor.

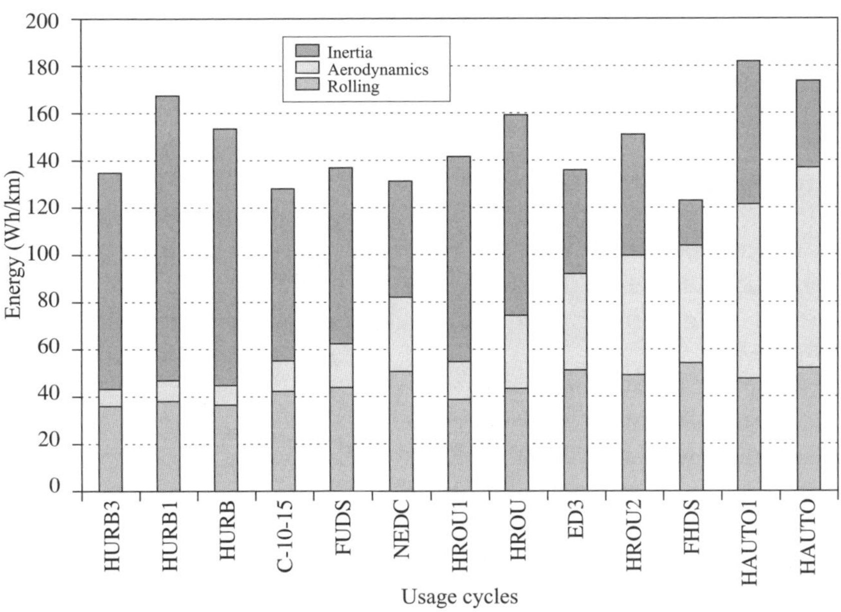

Figure 3.20. *Example of distribution of energy necessary to advance according to the type of use (1,600 kg private vehicle)*

The recovery of energy during braking is very important in urban-cycle uses, for which a great part of the traction is recoverable because of the low speeds (low dissipations in aerodynamics and the bearing) and frequent decelerations. For the example indicated in Figure 3.20, 40 to 50% of energy necessary for traction could be recovered in urban circulation conditions (without taking account, however, of the limitations of the drive train). This free contribution is important for the balance of energy flows of the buffer storage system and for the overall energy consumption.

This importance was illustrated in particular for heavy vehicles like urban public transport by bench tests on the North American series hybrid bus, Orion. The results, on a CBD cycle (20 kph mean velocity), indicate a value of 67.1 l/100 km for a reference NOVA diesel bus, a value of 54.6 l/100 for the hybrid (gain of 20%) and 63.5 l/100 km (gain of 5%) for the hybrid without recovery. It is noted that, under these conditions, most of the energy gain brought by hybridization is due to recovery during braking.

Implementation:

The implementation of recovery during braking can be delicate insofar as it interferes with the concepts of safety and driving comfort; we can distinguish the following main elements:

– an optimum recovery utilizes a first level with a relatively low recovery during foot-off-pedal decelerations, intended to restore the feeling of the engine brake then a second level with a greater recovery when the driver actuates the brake pedal (see Figure 3.21);

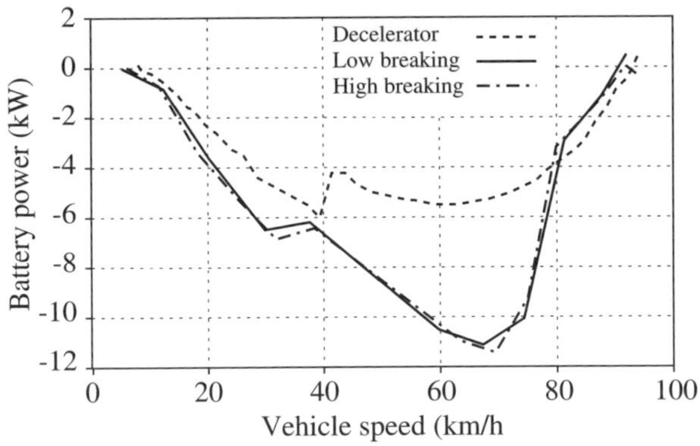

Figure 3.21. *Current measured on the battery without and with action on the brake pedal (Nissan TINO)*

– an optimum recovery during braking will require a fine management force sharing between the electric system, which will take the maximum, and the hydraulic system which will ensure the complement and will thus have to be controlled. Toyota proposes on the Estima an innovating system of hydraulic braking pressure regulation independent of the driver's action on the pedal, Electronically-Controlled Brake – ECB. In addition, Renault proposes on its concept car, Ellypse, a completely electric braking system provided by Delphi, Electro Mechanical Braking – EMB;

– the levels of power in deceleration are raised, inducing constraints on storage. An optimum recovery will require high-power systems like the super caps of associations of battery-super caps or even a flywheel;

– for parallel hybrids, the influence of architecture is important in that recovery cannot be optimum if the thermal engine remains connected to the electric motor during decelerations (configurations without all-electric mode). We will note on this last point, an interesting innovation on the Honda Civic hybrid for which a dynamic ordering of the valves makes it possible to isolate 3 of the 4 cylinders, thus decreasing by 50% the losses by friction of the thermal engine during decelerations;

Recovery is only assured on the drive wheels, a higher potential can be obtained with architectures with distributed motorization (see section 3.4.4.3.2).

3.4.4.2.6. Electric continuous transmission

Principle:

This configuration corresponds to a simplified hybridization of the series type, in which we have no reversible storage.

In the case of thermal-electric hybridizations we will speak of electric or diesel-electric drive, the latter having been well-known in the field of railway traction for many years.

As we will see this architecture was implemented in public transport applications.

This configuration has the advantage of not requiring a power buffer storage element; it presents the following disadvantages, however:

– the thermal engine and its generator must be dimensioned to ensure the totality of the traction power;

– the transmission does not make it possible to reduce the transients of power on the thermal engine, the entirety of the constraints on the power dynamics, related to the use of the vehicle, is deferred onto the generator;

– the vehicle does not have an all-electric mode;

– the vehicle cannot benefit from energy recovery during braking (except the case of electric auxiliaries which could be produced preferentially during these phases);

– the total efficiency of the electric drive (generator and traction motor) is lower than that of a mechanical drive in the majority of the uses.

However, this architecture shows the characteristics of a continuous transmission, with infinite amplitude, which is not the case for the mechanical solutions.

This solution makes it possible to choose, for each power required, the mode and the optimum torque of the thermal engine according to pollution, consumption or sound-level criteria (Stresses can exist, however, on the speed of the alternator enabling it to provide the power necessary to the correct performance of the electric motor).

Implementation:

The main advantage of the electric drive is the very big leeway that it allows in the layout of the components (important for public transport in particular).

As regards the operating conditions of the thermal engine, the advantages brought by the diesel-electric transmission to a traditional mechanical solution are illustrated in Figure 3.22.

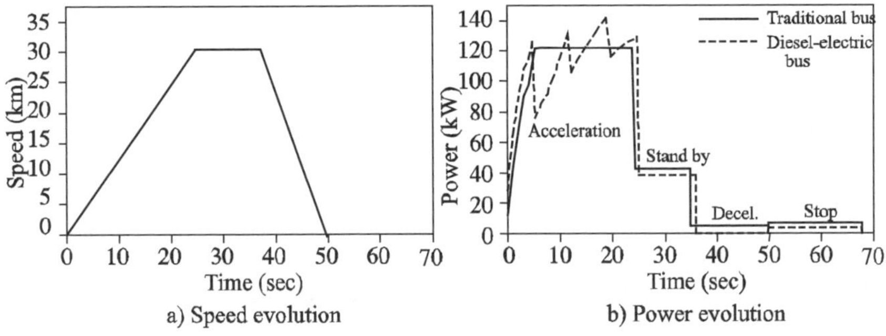

Figure 3.22. *Comparison of thermal engine output on a cycle Evolution of speed versus power*

For this simulation carried out within the framework of the V2G project by Renault VI, we compared the evolutions of the thermal engine output on a basic cycle of an urban bus.

In the acceleration phase, we can note the effect of the continuous transmission, which does not cause rupture due to the changes of gear. On the bearing we note in addition that the thermal engine, for both architectures, must adapt to the slack demand and is used at very low load.

The higher power observed on the thermal engine during this phase for the electric diesel is due to a lesser efficiency of the electric drive under these conditions.

We will see in the section on productions that this configuration does not make it possible, however, to reduce the fuel consumption of the vehicles.

3.4.4.2.7. Electric continuous transmission with storage (series hybrid)

Principle:

This architecture results from the preceding one by adding a component of reversible storage of energy; it makes it possible to have the improvements and the following functions:

– using a thermal engine of lower power, near the average power necessary, the supplements of power being covered by storage (see Figure 3.23);

– completely unlinking the operating conditions of the thermal engine and those imposed by the use of the vehicle;

– using the vehicle in all-electric mode with very good performance;

– recovering energy during braking.

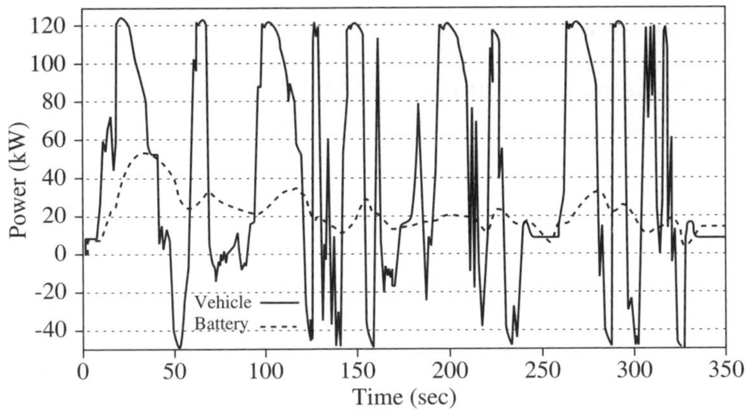

Figure 3.23. *Comparison of vehicle power and thermal engine power on a cycle*

Implementation:

As for the diesel-electric one, series architecture presents the disadvantage of a poorer overall efficiency and a significant number of engines and motors compared with the parallel architecture.

The series configuration is more particularly adapted to urban vehicles, used with low speeds and many stops; it then makes it possible to appreciably improve the operating conditions of the thermal engine and to recover a great quantity of energy during braking.

On the other hand, for extra-urban travel, the loss of efficiency on the transmission and the low gain on the thermal engine do not make it possible to achieve gains in consumption. Consequently, the applications of series hybridization on private vehicles, like the VERT project developed by Peugeot SA and Renault at the beginning of the 1990s, with gas turbines developing nearly 40 kW, were not continued further.

Range extender systems:

This configuration, is identical to the preceding one but, in this case, the power of the generator is so low that the vehicle can be regarded as an electric vehicle in which we have integrated an on-board charger. In fact the use of this vehicle is connected with that of an electric vehicle, the range extender only being put into action beyond a certain depth of discharge in order to cover a few additional kilometers (and if required to reassure the driver with regard to its autonomy).

This configuration was recently used by Renault which proposes its Kangoo in electric and hybrid versions with a range extender. The two versions use the same type of battery (NiCd 100 Ah, 13 kWh), the 12 kW generator of the hybrid version making it possible to prolong autonomy beyond the 60 to 80 km realizable in all-electric mode.

3.4.4.2.8. Gear-change assist

This additional service is intended to increase drivability and performance by deleting the effect of the interruption of torque during a gear change on a MGB or a RGB. Such a function is already proposed on certain mechanical drives using two clutches. In parallel hybrid architecture, this function can be obtained by using the torque of the electric motor in order to maintain the traction force during gear changes. This operation requires a particular configuration of the electric motor and gearbox as well as the use of a second clutch.

3.4.4.2.9. Damping of the angular oscillations

Mechanical solutions already exist for the absorption and smoothing of the torque oscillations which the alternative operation of the thermal engine can bring about. The fact of having an electric motor linked to the shaft of the thermal engine makes it possible to plan to fulfill this function by dynamic control of the torque damper function of the ISAD system. A precise assessment of the quantities of energies concerned in this operating process remains to be produced, however.

3.4.4.3. *Functional hybrids*

3.4.4.3.1. All-electric mode with autonomy

Principle:

Compared to the electric mode mentioned above, the all-electric mode with autonomy is characterized by the fact that we can force the vehicle energy management system not to activate the thermal engine. The vehicle can thus be used exclusively in electric mode over a given distance, this by decision of the driver, or directly via a positioning system or a signpost.

Implementation:

This operating mode, without local emissions of pollutants (vehicle said to be ZEV), is well adapted to European situations with very dense town-centers and relatively low speeds. The dynamic performance of the vehicle will thus be limited by that of its electric drive train, with autonomies of about 20 to 40 km.

We will note, however, that ZEV autonomy is also studied in large agglomerations affected by pollution phenomena linked to road transport, like Los Angeles. The stresses will then be stronger on the drive train, in terms of power due to a lot of expressways, and in terms of energy with required autonomies of more than 90 km, (typical autonomies in ZEV from 20 to 60 miles).

Electric autonomy will induce greater stresses on the battery, in-terms of cycling depth in particular; the on-board energies being about 4 to 18 kWh for applications in Europe or North America.

ZEV mode with autonomy will make it possible to add a value-adding function for a hybrid vehicle, in particular for the following cases:

– being able to go into a town center only authorized for vehicles not presenting local pollutant emissions. Such restrictive measures have been taken in the historical centers of some Italians cities (Milan in particular in Winter, 2002) and, more recently, London's congestion-charge, which includes a waiver clause for vehicles consuming alternative fuels [clean-auto.com 03]. It will be recognized, however that, for actions aiming at reducing urban congestion, the exemption granted to ZEV vehicles can only be conceived if they are very low in number; that nevertheless constitutes an advantage which can support their diffusion;

– being able to ensure the following missions with the same vehicle:

- delivery or collection in an urban center in all-electric mode, well-adapted to the low speeds and the many stops, over a relatively short distance;

- high-speed delivery, in hybrid mode with dispatching center office which can be far away from downtown;

- being able to enter confined spaces like depots, tunnels or mines;

– for military applications, being able to greatly reduce the thermal and/or acoustic signature in order to discreetly get as close as possible to a theater of operations. We can mention in this field the creation of the ECCE test platform in France (collaboration of DGA, Alstom and laboratories of North-Eastern France) or of freight vehicles in Great Britain.

3.4.4.3.2. Distributed motorization

Principle:

In this type of function, the hybrid motorization actuating an axle is supplemented by an electric motorization on the second axle.

Such a concept was proposed in particular by Renault back in 1996 on its Next with two 7 kW electric motors in the rear wheels.

Implementation:

This concept is particularly well adapted to the four-wheel drive market, for which it has the following advantages:

– allowing a fine distribution of the torque over the four wheels with management of anti-locking (ABS), anti-slipping (ASR) and differential sliding. The vehicle may be used as a 4x4 under the following conditions, (with increasing stresses on the battery):

- the boost function for acceleration requests;

- a 4x4 operation when requested by the driver, over a short distance for traction requirements;

- better stability of the vehicle on bends while benefiting from a recovery on the rear axle;

- ZEV mode with propulsion by the electric axle only;

- simplified mechanical architecture by removing the inter-bridge driveshaft in particular, which makes it possible to produce a 4x4 option with a minimum of modifications on the motorization and body compared to the initial version (Nissan Micra 4x4 minimum).

The development of hybridization in the 4x4 application could witness significant success because of simplifications and improvements made to services, in a sector where the price issue is less dominant. The Toyota Estima (E-four) vehicle, for example, has experienced great success in Japan, in spite of an additional cost of about 2,000 Euros compared to the traditional version.

3.4.4.3.3. External electric connections

Until recent years, the electric connection of a vehicle went as far as the network, and for a recharge of the battery, the most widespread case being the electric vehicle, but hybrid applications were carried out.

The recent events occurring in electricity distribution – or more precisely, non-distribution – in North America led the research centers to evaluate solutions in which the vehicle and its on-board energy storage would be connected in an active way to the network and could exchange energy in both directions.

In addition, car manufacturers, in a quest for valorization of hybridization, studied the possibility of supplying electric consumers from a vehicle.

These three options are presented below. Whilst many points still remain to be specified with regard to them, it will be advisable to follow their development because they may precede the future use of multifunction vehicles much more integrated into domestic and/or professional uses and for which transport itself will constitute only one of the possible applications.

Recharging by the network

Principle:

The management of the energy storage system of a hybrid vehicle can be assured in two ways, that is to say:

– maintaining the state of permanent charge around a median value (mode known as charge sustaining);

– decreasing the state of charge from its initial value up to a limit value controlled by the management system (mode known as charge depleting).

In charge-sustaining mode, the battery is never recharged from an external network, the vehicle only uses fuel and the driver can even be unaware that it is hybrid (it is the case of the first marketed hybrids).

In charge-depleting mode, it is necessary to recharge the battery periodically, at the end of the journey, or day, by the external network using a charger (on-board or not). The vehicle will then consume fuel and electricity, which will make it possible to ensure a transfer of primary energy consumption from oil towards other resources like natural or renewable gas via the electrical production mix.

Implementation:

Prototypes using the principle of network recharge were produced in the 1980s; we can mention the parallel hybrid vehicle of the DOE or hybrid VW Golf launched in 1983 (see section 3.2).

More recently, French manufacturer Gruau put its Microbus on the market, a public transport vehicle, the hybrid version of which uses a 11 kWh NiCd battery which must be recharged every evening at the depot.

We can make the following remarks for charge-depleting operation:

– the energy-pollution assessment cannot be carried out any more on the vehicle, it is necessary to carry out a complete assessment including the infrastructure of electrical production according to various assumptions on the production mix (geography, time, etc.);

– recharging on the network makes more sense with a high-energy battery (10 to 15 kWh); consequently it will be logical to use it linked with electric autonomy mode.

Active connection to the network

Principle:

This new concept, studied in particular in the United States, envisages energy exchanges in both directions with in particular a possible network supply by the vehicle's battery. In fact, the idea of the vehicle having an active connection to the network (concept of vehicle-to-grid or V2G) makes it possible to consider the following options:

– sale by a car owner to an operator of part of the energy contained in its battery at the highest price (consumption peak, rupture of the network, etc.);

– utilization of the vehicle in charge-sustaining mode, the autonomy being ensured by the fuel tank;

– utilization of the vehicle in charge-depleting mode with battery recharging during the night under favorable financial conditions;

– connection of the vehicle to a local network to contribute to the back-up supply of a hospital, or a police station, in the event of major outage of the network. Such a use can be envisaged with the requisition of a fleet of urban hybrid buses.

Implementation:

For a private consumer or a fleet, the charge and discharge criteria would be related to the immediate cost of energy, a concept which has a sense in a context of strong liberalization of the electricity market. It is a complementary function which could be brought to a private consumer, a fleet manager or to the community, however some points remain to validated, such as, for example:

– security management of the network when EV battery is used as an energy storage system on the network;

– cost of the corresponding infrastructure enabling sufficient flows of energy (transmission of power and information);

– overall cost for the owner, including the over consumption induced by the transportation of a heavier battery, and its replacement cost;

– such a concept only makes sense if there are a very great number of vehicles of this type on the road.

Such concepts are currently being studied in California by the EPRI and University of Davis, from a technical and economic point of view. Experiments are envisaged but there no marketing date is announced.

Supply of external auxiliaries:

The presence in a thermal-electric hybrid vehicle of a large battery, a high-voltage electric network and electronic converters makes it possible to consider the supply of external consumers in high voltage, thus adding a value-adding function to the vehicle.

This concept was proposed in particular by Citroen with a 220 V AC plug on the Dynalto project and in the United States with GM or DaimlerChrysler or Ford, who propose a 110 V AC plug on their pickup or hybridSUVs (power up to 20 kW available for the Dodge RAM).

Such a functionality makes it possible to supply professional or leisure equipment, for high powers; we can even imagine supplying an insulated house (energy being able to be provided for the duration by setting in motion of the thermal engine).

3.4.5. *Evaluation of hybrid vehicles*

3.4.5.1. *Issues*

The thermal-electric hybrid vehicles include a thermal engine and can be compared with traditional vehicles in terms of both their dynamic performance and their autonomy. Consequently, these vehicles may be tested according to the same procedures as their thermal counterparts, in particular for determining their regulated atmospheric pollution emissions (CO, HC, NOx and particulates for diesels) and their fuel consumption (and CO_2 emission). Some specific hybrid characteristics (detailed in Chapter 3) will have be taken into account during the bench tests, however; these points will relate to the vehicle, its energy management and the bench itself. We can mention:

– the operation of the internal energy storage, that is to say:

- load level variations between the beginning and the end of the test which can modify the results to a greater or lesser extent;

- the possibility of carrying out a recharge by the electrical network (grid charging);

– the vehicle energy management operation with the possibility for the driver to choose a ZEV operating mode with autonomy;

– the roller-bench operation, that is to say:

- energy recovery on the drive wheels during decelerations which cannot be correctly reproduced, the rear wheels being immobilized;

- the possibility of energy recovery in deceleration on the four wheels, which must be taken into account.

Some of these points will have an influence on the type of results, as we indicated in Table 3.4 for which we made the following assumptions:

– the vehicle with range extender can be recharged on the network;

– the vehicle with ZEV autonomy can have the possibility or not to recharge on the network;

– the vehicle without ZEV autonomy does not have the possibility to recharge on the network.

These assumptions are justified by the fact that the benefit of battery recharging on the network increases with the energy that it contains.

Architecture type	Electric	With range extender	Hybrid	Without ZEU autonomy	Thermal
Management	none	choice made in APU market	with ZEU autonomy	none	
Connection		recharge network	choice of ZEU mode	no recharge network	
	kWh/100	l/100 + kWh/100	l/100		

Table 3.4. *Classification of the types of vehicles and possible operating conditions according to electric and thermal vehicles are indicated for memory*

These various characteristics are detailed in the following chapters.

3.4.5.2. *Reversible storage of energy not rechargeable on network*

As we will see in Chapter 6, it is the case of the first vehicles put on the market which hold a quantity of on-board energy of about 1 to 2 kWh in their power batteries. Even if this energy is used only very partially, the quantity consumed is not negligible in relation to the energy put into play to drive the vehicle on the test cycle, about 1.5 to 2 kWh for 1,450 kg vehicle on the European cycle. The expression of vehicles' consumption and emissions must be made at constant-storage energy level in order to be able to authorize comparisons with conventional vehicles. Several solutions can be implemented in order to solve this problem, that is to say:

– linking up a sufficient number of cycles in order for the variations in the quantity of energy contained in battery storage to be made negligible in relation to the energy put into play during the cycles. This method, whilst it can give satisfactory results, presents the disadvantage of considerably lengthening the duration of the tests and making them depend on the quantity of energy contained in storage (for the same required precision). In addition, this procedure cannot apply to cycles with a cold start;

– bringing back the storage system to its initial load level by continuing the standardized test by specific evolutions, which are to be defined for each vehicle (operation when stationary, speed stabilized, etc.). This method presents the disadvantage of being dependent on the vehicle, and requiring information on the part of the manufacturer. It will be noted that if the distance covered in this second phase is not null, the question is asked as to taking it into account in the consumption and emissions calculation. The European standard for the measurement of energy consumption for hybrid vehicles specifies not to take this distance into consideration;

– carrying out several tests with various relative charge variations of the storage system with an aim of establishing a relation linking the values measured to the storage energy variations. These variations can be obtained by modifying the initial charge conditions of the storage system by specific tests.

This last method has generally been adopted by manufacturers and laboratories as presenting a compromise of precision, duration and adaptation at the various cycles. Figure 3.24 illustrates the variations in the depth of discharge, in relative values, on the battery of the hybrid Nissan Tino during a series of tests on the European cycle with various initial conditions.

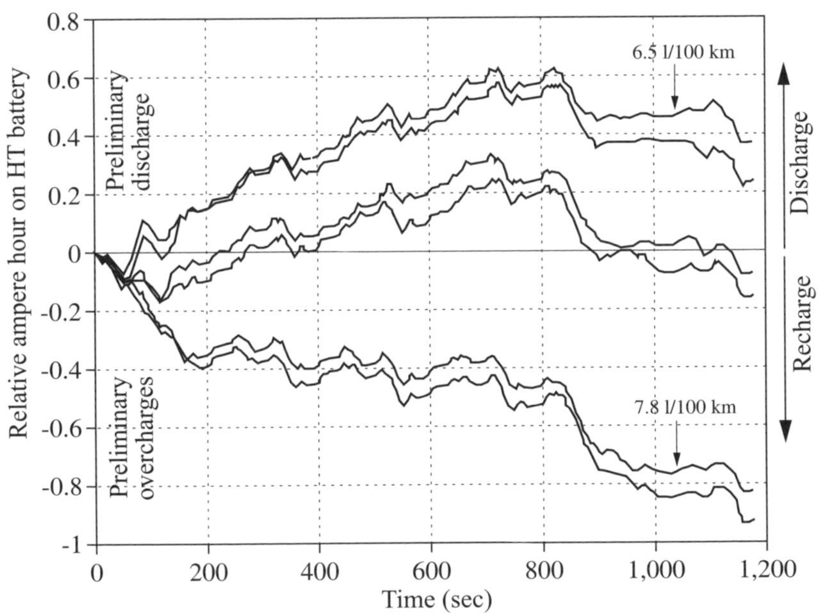

Figure 3.24. *Example of relative variations of the SOC battery for a hybrid Nissan Tino*

If we consider the variations of energy consumption according to the depth of discharge, it would seem that a linear expression can be used in virtually all the situations with a good precision.

This result translates the fact that the overall energetic efficiencies vary little on each drive train.

It will be noted that it has been possible to highlight appreciably nonlinear variations for an extreme situation of a thermal engine used in series on two operating points presenting very different efficiency.

For the vehicles tested so far, a linear interpolation has been produced in order to express the fuel consumption corresponding to zero variation in the energy storage load level (see Figure 3.25).

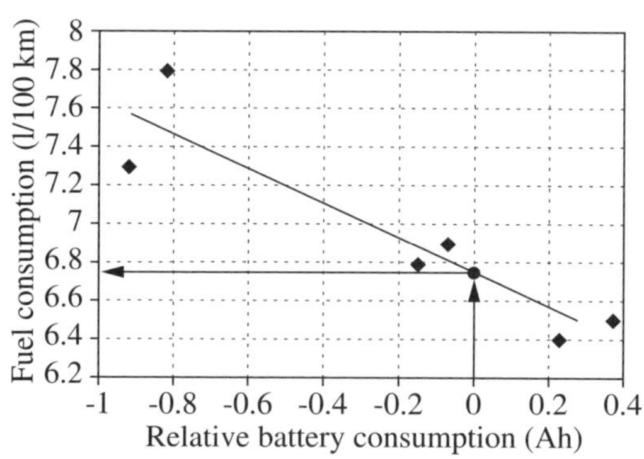

Figure 3.25. *Expression of consumption according to the battery charge for a hybrid vehicle*

The coefficients of regression obtained are high, generally above 0.9. Lower values can be obtained in the case of cycles with cold start, because of a generally lower number of measurements and a greater dispersion related to the thermal transient conditions.

Some authors have proposed, under these conditions, to separate the cycle into several under-cycles that are more homogenous in thermal conditions and thus presenting higher coefficients (about 0.9 or more compared to 0.75 overall).

Such a method makes it possible, for example, to increase the overall coefficient from 0.75 to 0.92; we can note, however, that this distribution does not lead to significant variations of the value of consumption obtained (less than 1.5%).

The variations in the battery charge level are expressed using an integration of the current, given in Ah, which implicitly supposes that the faradic efficiency of the battery is one, therefore implicitly a non-deteriorated battery used in the operating range initially envisaged. In addition, the integration of the current will require great attention on the sensor (precision, offset, etc.).

However, the experiments carried out in various laboratories show that the emissions do not vary in a coherent way with the storage state of charge, consequently the preceding method cannot be used and the values are calculated by carrying out a simple averaging of the measurements.

3.4.5.3. *Reversible storage of energy rechargeable on network*

Some hybrid vehicles authorize a recharging of the battery on the network (charge depleting), which makes it possible to carry out a transfer of energy consumption from a fossil fuel to electricity. In this case the procedure requires the fuel electricity consumption to be measured, and a total assessment to be carried out (pollutants and GHG) including the production of the two energy vectors used. For electricity it is necessary to take into account the production mix, which influences in an important way the GHG emissions. Table 3.5 indicates some values suggested for various scales in space and time.

Whilst for regulated pollutants we can imagine considering a national situation, for GHGs it appears relevant to take into account data on a broader basis like

Europe. In any event a specific procedure will have to be defined with the public authorities in order to define and to standardize the values to be taken into account and the calculation making it possible to compare these vehicles with their traditional counterparts.

CO_2 (g/kWhe)	1995	1999	2010	2020	2030	2050
Europe	468		446		384	307
France	67		67		103	141
France			120			
Germany			563			
Denmark			803			
USA according to EIA		710		757		
USA according to CEF		710		520		

Table 3.5. *Emissions of CO_2 induced by the production of electricity according to the countries or group of countries and future predictions*

3.4.5.4. *Taking into account the electric mode with autonomy*

If the electric mode proves best adapted to the speed curve, it may then be selected, as standard NF EN 1986-2 envisages. In this case it will be necessary to make sure that the thermal engine is put into action during the test. The procedure thus envisages that the total distance corresponding to autonomy in electric mode is exceeded during the test (the cycle in progress must be finished, immediately followed by a new complete cycle). With such an operating procedure we can be certain that the thermal engine will be started, in cold conditions, during the test procedure.

Once the test is finished, the taking into account of the battery state of charge variations will depend on its connection, possible or not possible, on the network; we will refer then to sections 3.4.5.2 or 3.4.5.3.

3.4.5.5. *Taking into account of braking on roller test bench*

A vehicle has a system of distribution of the braking force between the front and back axles, which makes it possible to respect the fundamental criterion of safety, which is never to block the rear wheels before the front wheels. This system will consequently favor braking on the front axles, all the more so as the braking force required is considerable. For two–wheel-drive vehicles, which are the large majority of the market today, the roller bench tests were carried out using one (or two) roller(s) slowing down the drive wheels. Under these conditions, the non-drive wheels being immobilized, the driver only has braking on the drive wheels to follow the desired speed profile. For vehicles authorizing energy recovery on the drive wheels, it is necessary to check that this operating process does not over-estimate the quantity of energy returned to the battery. Such experimentation was carried out at INRETS during the first tests on electric vehicles. An electric Citroën AX, provided by Peugeot SA and equipped with sensors was tested on a road, without slope, then on a roller test bench according to the same profile speed (recorded). The electrical energy results, measured at the battery, reveal that the over-estimation on the roller test bench does not exceed 4% on average over the ten tests carried out. The electric vehicle has a relatively low capacity for energy recovery, however; comparisons on the roller test bench have since been carried out on hybrid vehicles. The JARI carried out comparisons on two hybrid vehicles with decelerations varying from 1 to 2 m/s^2; the results showed that the currents recovered on the battery were comparable.

3.4.5.6. *Distributed braking*

As indicated in section 3.4.4.3.2, hybridization is a means of having a distributed motorization and is in fact applied by Japanese (Toyota Estima) or American (Ford Escape) manufacturers. The testing of such vehicles then requires a 4-wheel-drive bench or a special procedure which must be put in place with the collaboration of the manufacturer. Under these conditions it is then possible to ensure that the main electric motor recovers the totality of the energy which would have been recovered on the two axles in road use (if the engine has the sufficient capacities). Such a procedure was notably used for the certification of the Toyota Estima. If the conditions above are not met, however, a 4-wheel-drive bench will have to be used.

3.4.6. *The first vehicles on the market*

3.4.6.1. *Passenger cars*

3.4.6.1.1. Which hybridizations?

The great number of architectures and the variety of components we can use no doubt explain why many prototypes were constructed by manufacturers and suppliers. A study carried out by the Hybrid Vehicles working group of the International Energy Agency put their number at more than 60; the reader will be able to refer to it for an exhaustive knowledge of these productions.

Among these productions, very few of these concepts have been proposed on the market at the present time, and if we consider the richness of the technological solutions and the functionalities which have been studied, it would seem that the first marketed vehicles are very much in decline. Such a situation can be justified in several ways; we can mention:

– limiting the additional manufacturing and maintenance costs for manufacturers placed in a context of free competition new and second hand vehicle market;

– not depending on the need for a specific infrastructure, like an electric recharge connection, this being one of the difficult factors in the diffusion of electric vehicles. Manufacturers thus seek to differentiate their hybrids from the electric vehicle. Indeed, the information on hybrids published by Ford in 2004 is named “Unplugged”;

– not asking the driver to perform a specific intervention in vehicle energy management, which is entirely automated;

– depending on restriction measures for the use of traditional vehicles, which would make it possible to add value to functional hybrids. The implementation of such measures comes under the domain of the public authorities, who at the present time remain quite reticent (except for some examples in Italy and London and a willingness in Southern California).

Consequently, the first marketed hybrid vehicles are primarily of discrete type, their operating characteristics of not being easily detectable for the driver, their passengers or outside. Only an informed ear will be able to detect a thermal engine operation, which does not seem completely correlated with the dynamic evolutions of the vehicle. These first hybrids consume conventional fuel, no outer electric connection for the battery, and their driver does not have any more control latitude than in the case of an automatic transmission.

Such an evolution can be illustrated by considering the size variations of the vehicle battery pack; which is indeed given as one of the main factors of the additional purchase cost (operating life, resale). Figure 3.26 presents battery energies according to the possible ZEV autonomies for a set of productions.

We note a large variety in capacity in electric mode proposed on the prototypes (with their incidence on battery energy) and the particular conservative situation of vehicles marketed with zero autonomy, which enables them to be satisfied with batteries of very low dimension, the NiMH solution seeming to be the first mature one on the hybrids market.

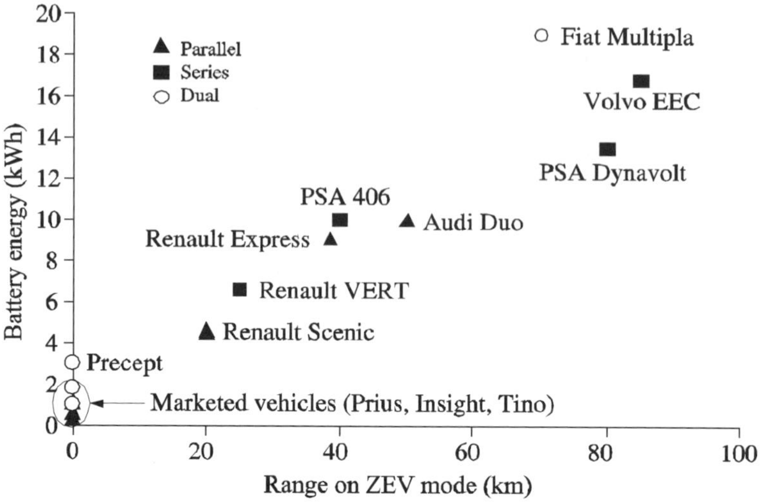

Figure 3.26. *Evolution of ZEV capacities (prototype and marketed)*

The functions brought by the various creations are indicated in Appendix 3 and the main characteristics in Appendix 4.

3.4.6.1.2. What performance?

The first vehicles have been marketed almost exclusively by the Japanese, with gasoline motorizations. These pioneers present very good consumption performance, particularly in urban conditions, their emission levels are those of the future standards and their dynamic performance is equivalent to that of traditional vehicles.

In terms of greenhouse gas emission, these vehicles are better placed than their direct competitors, diesel motorizations.

If the purchase price and the energy cost are considered, the first hybrids are on a par with diesel motorizations.

We can, however, make the following remarks:

– comparisons in terms of emissions must take into account the fact that diesel motorizations present particulate and NOx emissions which remain high and require the use of filters and specific catalysts, which are not yet proposed on all vehicles;

– comparisons in term of cost takes into account purchase subsidies (€1,525 in France, $1,500 in the USA) and differentials on fuel cost (in Europe particularly). The future variations of these factors may influence the competitiveness significantly highlighting the political decisions.

3.4.6.1.3. What strategies?

If we consider the technologies implemented, we note that the first manufacturers diffused very varied architectures, from the non-integrated alterno-starter to the distributed motorization.

This large diversity would seem to illustrate the extent of the questions that manufacturers and suppliers ask when it comes to the choice of a solution with respect to the external parameters (competition, regulations, policies, energies, etc.).

This situation seems to continue beyond the first vehicles, if we consider, just for 2004, the marketing of the Ford Escape with its distributed motorization, and that envisaged by the PSA group of the Citroën C3 with a non-integrated alterno-starter.

If we consider the industrial strategies implemented, few elements are diffused but we can distinguish some situations, such as:

– the research for integrating technologies used in hybrid vehicles, like Toyota developed for its Prius (see Figure 3.27 below);

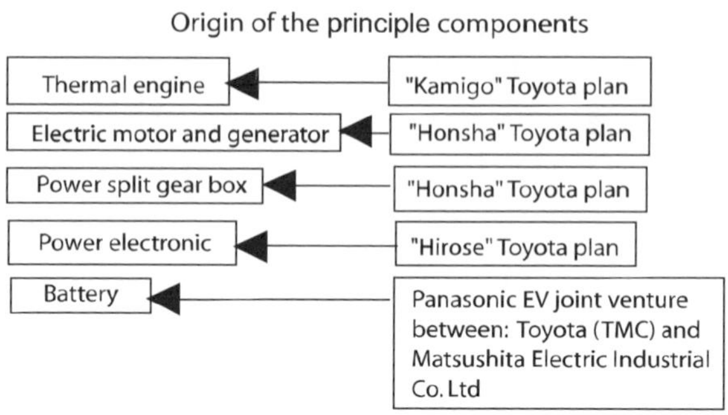

Figure 3.27. *Toyota production organization for the Prius (Toyota document)*

– the purchase of licenses from more advanced manufacturers, as is the case for Ford and Nissan with respect to Toyota;

– the integration of subsystems fulfilling some functions in discrete hybrids, as is the case for Citroen with Valéo on the C3 or Renault-Nissan with Bosch and Yazaki on the Modus.

If we consider the guarantees brought to the customer, it seems that a consensus is appearing among the first manufacturers concerning a warrantee for the hybrid system, or its battery, covering eight years and/or 160,000 km (100,000 miles).

3.4.6.1.4. The Toyota Prius

The engine:

From a technological point of view this vehicle is a success since it implements, successfully, a parallel–series type of architecture (parallel with power split).

This type of coupling dates back some time; it was already proposed at the beginning of the 1980s by English firm Lucas in a prototype.

Taking into account the limited performance of electric motors and their control, this prototype, perhaps too advanced for its time, did not undergo any later developments.

The parallel-series concept was taken up again by the Japanese companies "Equos Research" and "Aisin", who presented an architecture at the end of 1996 that was very close to that retained by Toyota.

With the Prius hybrid vehicle having been much publicized in the media, its motorization is now well-known (see diagram in Appendix 5); we will insist, however, on its global optimization performance, with in particular the following objectives:

– only use the thermal engine in the phases where its operation is necessary: for the dynamic performance, the drivability or management of the battery or the thermal transients. In the opposite case the thermal engine is stopped and the vehicle is operated in all-electric mode;

– when the thermal engine is used, make it function in its zone of better efficiency, which is obtained in particular by the management of the parallel-series energy flows.

The first level of optimization can be illustrated by a fine analysis of the operation, which simulation makes it possible to highlight, after validation on bench tests.

Figures 3.28 and 3.29 present two cases of vehicle operation. One in congested use at very low speed, and the other in more fluid urban use.

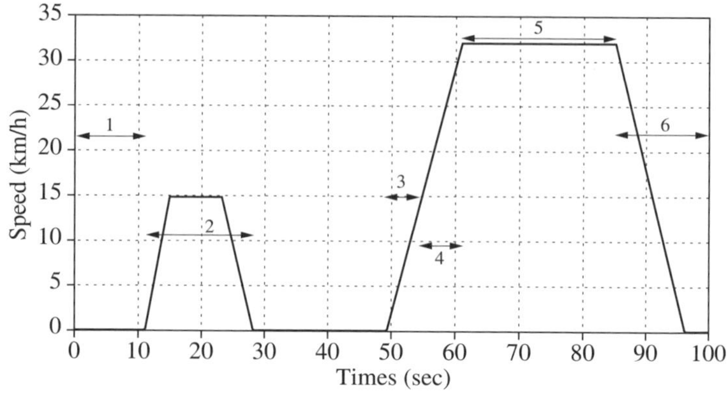

Figure 3.28. *Evolution of vehicle speed*

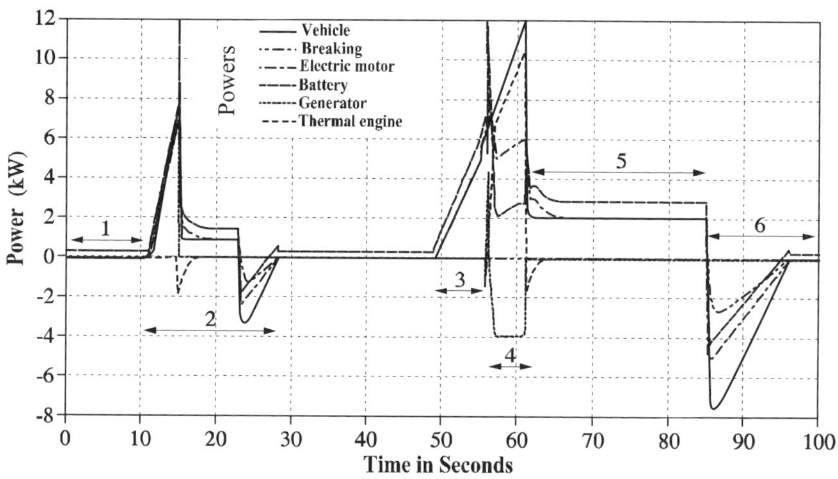

Figure 3.29. *Evolution of power in the different components*

For a vehicle moving with its thermal engine at hot temperature and on a flat road, the following operating phases can be highlighted:

– when the vehicle is stationary the thermal engine is turned off; we will note that the air-conditioner compressor is driven electrically on the 2003 version;

– when the vehicle is used at very low speed the thermal engine is not put into action and the vehicle moves in all-electric mode propelled by the electric motor, and the generator is not in operation;

– when the vehicle starts, setting-in-motion occurs in the all-electric mode, thus removing the clutch skidding phase, not very efficient, that exists on conventional motorizations;

– when the vehicle accelerates, the thermal engine is put into action and we observe the series flow in hybrid mode. Indeed, the thermal engine provides energy directly onto the shaft but it also actuates the generator which provides electrical energy on the network. The electric traction motor, which assists the thermal engine in this phase, will use electrical energy coming from the generator and the battery via the power network. This series operating process explains why the vehicle can have an electric motor providing 50 kW on the shaft whereas the battery can provide only a maximum of 38 kW of electricity;

– when the power demand falls, at stabilized speed in our example, and when the vehicle runs at low speed (maximum limit near to 50 kph on a flat surface) the thermal engine can be stopped. We thus return to the all-electric operation considered previously;

– when the vehicle decelerates, the thermal engine is stopped and all the recovery authorized by the system can then be carried out.

The second level of optimization is obtained in particular by the use of series and parallel flows.

It is observed indeed that, in the bearings at mean speed where the thermal engine is put into action mainly for drivability needs, its torque is increased because a recharge of the battery by the electric motors is carried out.

The thermal engine thus ensures the propulsion of the vehicle in parallel mode with a good efficiency and the maintenance of the battery state of charge by the series flow, which makes it possible to increase its efficiency.

Such an operation thus makes it possible to very strongly limit the uses of the thermal engine to very low torque, frequent in urban use, as is illustrated by Figure 3.30, where we have represented the evolution of the operating points of the thermal engine for two urban uses.

If we compare the zones of use with those indicated for a conventional motorization (see Figure 3.10) we clearly note the optimization of the range of use in urban conditions.

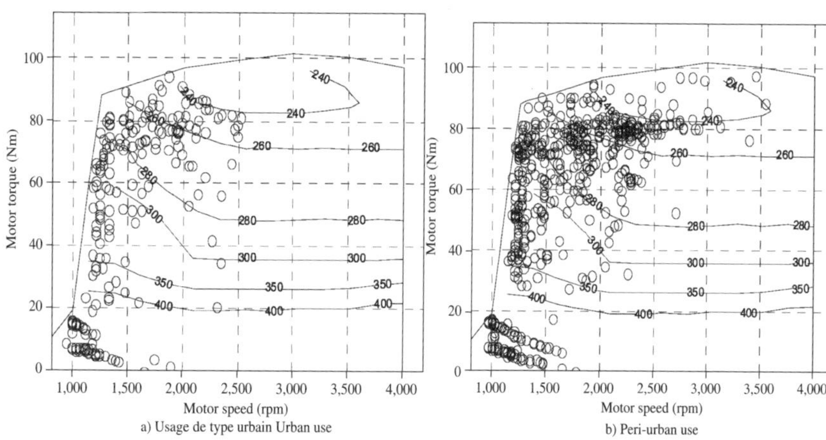

Figure 3.30. *Evolution of thermal engine operating points of the Prius 1997 (measurement and simulation: INRETS-IFP)*

These observations on the operation of the Prius are very much consistent with the measurements carried out at INRETS in collaboration with the IFP, on the first model.

Indeed, the comparisons carried out with vehicles with a conventional gasoline motorization indicate that the average gain in consumption is at least 30% in urban conditions, and 15% on roads and not appreciable on motorways.

We will notice that the operating processes detailed in this chapter can be modified, in particular according to:

– the temperature of the thermal engine, with a specific procedure being used during the engine heating phase and it post-treatment;

– the initial conditions of the battery (state of charge, temperature, etc.);

– the operation of the auxiliaries (air-conditioning);

– the activation of the all-electric mode present on the 2003 Europe version of the vehicle.

The vehicle:

Toyota achieved a first by announcing the marketing of the first hybrid vehicle at the EVS14 Conference in November 1997 in Orlando.

Since this date, Toyota has already created two new, improved of its vehicle and sold more than 600,000 units in Japan, North America and Europe.

In the case of Europe, the sales objective remains modest with 5,000 vehicles for 2004, and 15,000 today, Toyota proposing a warranty covering eight years or 160,000 km on components specific to hybrid motorization.

For each new version, the performance of the vehicle was improved, as is illustrated in Table 3.6.

Specifications	Model 1997	Model 2001	Model 2003
Emission on cycle CEE in g CO_2/km	136	120	104
Total equivalent power in kW	64	74	82
Performance of 0–100 km/h in s	15	12.9	10.9
Weight in kg	1,250	1,270	1,300
Max power of thermal engine in kW and rpm	43 @ 4,000	53 @ 4,500	57 @ 5,000
Max torque of thermal engine in Nm and rpm	102 @ 4,000	115 @ 4,200	115 @ 4,200
Max power of electric motor in Nm and rpm	30 @ 940	33 @ 1,040	50 @ 1,200
Max torque of electric motor in Nm and rpm	305 @ 940	350 @ 400	400 @ 1,200
Max voltage V	275	275	500
Battery	Panasonic NiMH cylindric	Panasonic NiMH prismatic	Panasonic NiMH prismatic
Nominal voltage of battery pack in V	288	274	200
Max power of battery pack in kW	35	40	38
Total weight of monoblock in kg	44	40	29
Specific power in W/kg	800	1,000	1,300
Specific energy in Wh/km	40	46	46

Table 3.6. *Characteristic compared on three Toyota Prius models*
**: 10 s, 50% SOC, 25°C*

We will notice the increase in the maximum power delivered by the thermal engine and the electric motor from one version to the next, with consequently an improvement to the dynamic performance compared to the 1997 version, which did not perform very well on this point. We will also note the progress made on the battery with a very significant increase in the specific power. The use of a booster makes it possible in the latest version to reduce the number of battery elements while making the electric motors work with a voltage able to reach 500 V.

The selling price of this vehicle in France is €23,500 (from ADME (French environmental agency) tax credit of €2,000 deducted), the comparison with its competitors equipped with a conventional motorization is not easy because of a great number of parameters to be taken into account. Table 3.7 makes it possible to compare Prius with two close models comprising gasoline and diesel motorization.

Vehicle	Prius 2003	Renault Laguna		Peugeot 407	
Engine	Petrol 1.4	Diesel 1.9 DCi	Essence 1.6 16V	Diesel 1.6Hdi	Petrol 1.8 i
Thermal engine power, kW	81	79	79	80	85
Weight in kg	1,300	1,350	1,270	XX	1,400
Size Lxlxh in m	4.45x1.73 1.49	4.56x1.77 1.43	4.58x1.77 1.43	4.67x1.81 1.44	4.67x1.81 1.44
Volume of trunk in l	408	430	430	430	430
Acceleration 0 to 100 Km/h in s	10.9	12.3	11.5	11.7	11.3
Emissions based on the MVEG cycle	104	152	173	145	187
Consumption based on the MVEG cycle in 1/100 km	4.3	5.6	7.2	5.5	7.9
Energy cost* €/100 km	4.56	4.70	7.63	4.62	8.37
Price of car in€	23500**	24100	21400	#21500	#19400

Table 3.7. *Characteristic compared on several models close to the 2003 Prius *: Hypothesis, gasoline €1.06 / l and diesel €0.84 / l*
***: ADEME taxcredit deducted = €2,000*

Based on these values we can make the following remarks:

– if we consider CO_2 emissions, the Prius is very advanced, with a gain of almost 30% with respect to the nearest vehicle;

– if the dynamic performance is considered, it appears that the Prius is best placed because of the contribution of electricity in the motorization. However, this result is to be tempered if we consider performances over longer durations (mountain, towing) because the contribution of the battery cannot then be assured anymore (we will then find performance close to that of the other vehicles because of thermal engine power ratios on neighboring mass);

– if we consider the costs of purchase (reduction deducted) and energy, the Prius is comparable to the two diesel models, with the gasoline-diesel differential scenario existing in 2004).

3.4.6.1.5. The Honda Insight

Motorization:

This vehicle has an integrated alterno-starter-type booster architecture and its operating processes correspond completely to what was presented in sections 3.4.4.2.1 and 3.4.4.2.2. It will be noted that this vehicle has an integrated electric motor of which the thickness does not exceed 6 cm for a power of 10 kW.

The battery type is NiMH from Panasonic EV in cylindrical version, similar to that of the first Prius.

The vehicle:

This vehicle was marketed by Honda in November 1999 in Japan then in North America and Europe since 2000. Arriving on the market after Toyota, Honda engineers have sought to assert their presence by the consumption of their vehicle; with an objective to reach 3 l/100 km on the Japanese cycle (with Toyota announcing 3.6 l/100 km). To this end an approach of total synthesis was carried out, intervening in particular on the following three axes:

– the vehicle, with two seats only, high aerodynamic performance (Cx 0.25) and an aluminum body (empty weight 820 kg);

– the 3-cylinders thermal engine, in particular associating optimized control of the 16 valves (VTEC-E), with reduced friction, optimized internal aerodynamics and lean burn mixture;

– hybrid transmission (IMA) with 10 kW integrated electric motor (MGB or CVT version).

Honda thus achieved its goal of a vehicle with 3 l/100 km (Honda data: 2.9 l/100 with manual gearbox, results of the ADEME EVALVH program 3.1 l/100 with CVT). According to the manufacturer, the energy gain is distributed equally over the three optimized parameters. In addition the vehicle presents very low emission levels and respects the European standard Euro 4.

It will be noticed that this vehicle, with two seats only for a length of 2.9 m, optimized aerodynamics and a very low body is not particularly optimized for urban use, where the discounted gains are, however, highest. We can think that Honda thus created a technological window, difficult to classify, and of limited diffusion, allowing technologies to be validated before establishing them on more popular models (see below).

3.4.6.1.6. The Honda Civic

This widely-diffused vehicle takes up the IMA system. It was put on the French market in December 2003, when more than 15,000 vehicles had already been marketed in Japan and the US. Like the Insight, this vehicle is proposed with a 5-speed manual gearbox or a continuous transmission, only the manual gearbox being proposed in France. The positioning of this hybrid is in principle easier than for the Prius because it uses an existing platform. However, the Honda marketing policy means that the four-door model, the basis of the IMA, is not exported in Europe. We have therefore compared it with the closest gasoline and diesel versions, which are five-door and manufactured in Great Britain (see Table 3.8).

Vehicle	4 doors IMA	5 doors 1.4i LS	5 doors 1.7 LS CDTi
Engine	Petrol IMA	Petrol 1.4 i	Diesel 1.7
Power Thermal engine, kW	61XX	66	74
Weight in kg	1185	1125	1304
Acceleration 0 to 100 Km/h in s	12.8-10,5 XX	11.9	11.5
Emissions based on CEE cycle in g CO2/km	116	153	134
Consumption based on CEE cycle in l/100 km	4.9	6.4	5.0
Energy cost €/100km*	5.20	6.78	4.20
Price of car in €**	20400**	16800	20400

Table 3.8. *Characteristics compared on three Honda Civic models*
**: Hypothesis, gasoline €1.06 / l and diesel €0.84 € / l*
***: ADEME tax credit deducted = €1,525.*

Honda proposes for the IMA Civic a guarantee of eight years and/or 160,000 km on the battery (the most sensitive component).

The comparison of data on the costs of purchase and energy between the IMA and the diesel version are close to those indicated for the Prius, the hybrid being slightly disadvantaged here with regard to energy cost. Indeed the rather simple hybrid architecture of the IMA does not lead to a gain in consumption making it possible to make up for the gasoline-diesel cost differential.

3.4.6.1.7. The Toyota Estima

In this vehicle, Toyota has further improved its hybrid concept (THS-C) by adapting it to the four-wheel drive motorization. A THS motorization, near to that of the Prius, but incorporating a CVT, is used on the front axle whereas a 18 kW electric motor drives the rear wheels (see Figure 3.31).

An optimized management of the overall vehicle system makes it possible to have all of the distributed motorization functions which we developed in the preceding chapters, this both in acceleration a in regenerative braking and whatever the level of adherence to the road.

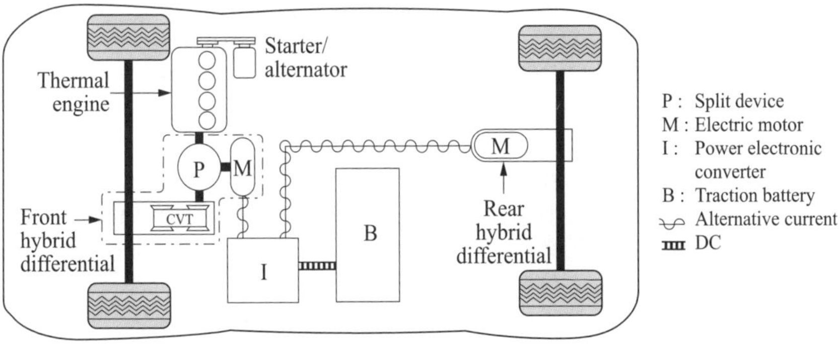

Figure 3.31. *General diagram of THS-C or E-Four transmission*

For this vehicle, Toyota announces a consumption of 5.6 l/100 km on the Japanese cycle (10-15 modes), compared with 11 l/100 km of the initial version. This cycle proves very favorable to the hybrid concept and we find the level of the gains announced for Prius.

It will be noted that this vehicle is the first marketed in large volume with a fully electric brake system.

This architecture undoubtedly represents the highest degree of sophistication which we can find to date in hybrid motorization.

The Estima hybrid has been marketed on the Japanese market since the beginning of 2001. In 2002 Toyota had sold more than 12,000 Estima units; the additional purchase cost is about €2,000.

3.4.6.1.8. The Ford Escape hybrid

This hybrid vehicle was announced by Ford in April 2004 at the New York Motor Show.

The architecture chosen by Ford is similar to that of the Toyota Prius, it uses a 4-cylinder gasoline engine (2.3l Duratec, 100 kW, Atkinson cycle), an epicyclical gear with a generator (called eCVT by Ford) and a NiMH battery (Sanyo, 330 V, cylindrical modules, size D).

The exact relationship of the technologies used in this vehicle with respect to those of Toyota is not known to date. Ford did indeed buy licenses from Toyota concerning the hybrids but claims to have developed and produced the motorization without integrating Toyota technology.

The vehicle is proposed in two- or four-wheel-drive version; in the latter case the architecture is undoubtedly close to that of the hybrid Toyota Estima.

The consumption announced by Ford is about 6.5 to 8 l/100 (EPA City and Highway), compared with 10 to 13 l/100 for the thermal version; the gain is thus significant (the potential on a SUV with V6 engine being significant).

Ford markets the vehicle at $26,400 for the two-wheel-drive version and $28,000 for the 4x4 version, the difference with the equivalent thermal vehicle being around $3,400 (tax credit of $1,500 to be deducted).

3.4.6.1.9. The Toyota Crown

Contrary to its other achievements, on the Crown, Toyota uses a very simple architecture of the non-integrated *stop&go* type with a 36 V lead-acid battery (THS-M, see Figure 3.32).

This system allows a gain in consumption on the Japanese cycle. Compared to the simple *stop&go*, the THS-M system proposes two innovations, that is to say:

– the use of a clutch allowing the thermal engine to be untorqued from the pulley-belt system binding the electric motor and the air-conditioning compressor. The opening of this clutch during stopping makes it possible to ensure the drive of the compressor and maintaining of the air-conditioning (see section 3.4.4.2.1);

– the setting in motion of the vehicle by the electric motor, before activating the thermal engine. This function is of interest with regard to energy, but it will undoubtedly be limited on this vehicle of more than 1.6 t by the 3kW motor power.

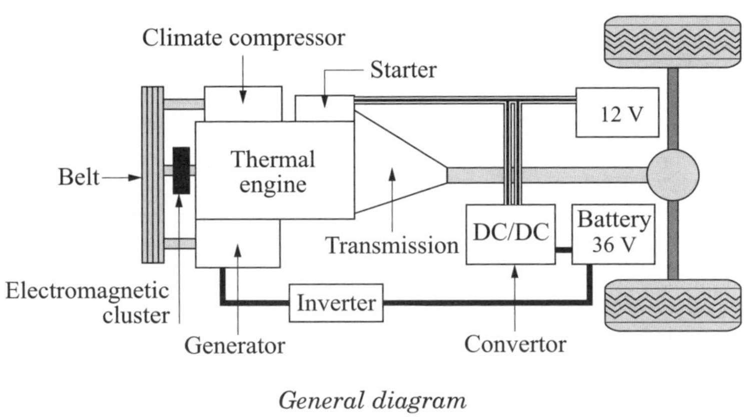

General diagram

Implementation

Figure 3.32. *Motorization of the Toyota Crown hybrid (Toyota document)*

3.4.6.2. *Public transport*

3.4.6.2.1. Issues

In the particular case of urban public transport, research concerning innovating transmissions is justified by slightly different criteria from those which we indicated for private vehicles; we can mention:

– reducing or eliminating the local gaseous pollutant emissions, this point being of most importance for vehicles with exclusively urban use;

– reducing the energy consumption, which is important for the owner and more generally for the greenhouse effect if fossil fuels are involved;

– improving accessibility for passengers, in particular when getting in and out and during displacements in the vehicles. This function can be obtained with the installation of a low floor, if possible over the entire length of the vehicle (integral);

– giving more flexibility in vehicle energy supply. That can result in partial suppression of an overhead line, or more generally looking for cost management with the use of generic components (same electric drive train for buses, trolley buses or series hybrids).

When the various achievements are considered, it appears that the order of taking into account of these criteria can be variable, bus manufacturers having implemented several types of hybridization. The examples quoted hereafter are only indicative; for a detailed list of the various projects, refer in particular to the work of the IEA Hybrid Vehicles working group.

3.4.6.2.2. Diesel-electric implementations

Irisbus

This solution was studied with the aim of producing a low floor (320 mm) on the entire length of the vehicle whilst in addition creating the possibility of a maximum-width circulation corridor between the drive wheels. This realization was done in France, within the framework of collaboration between three companies, with the funding of the public authorities. The solution is based on an asynchronous motor with its two-stage reducer (Alstom Transport) which was able to be housed in the wheel thanks to the development of an extra-wide tire (Michelin) and integration with a mechanical oil-bath brake, produced out by "Renault Véhicules Industriels" – now Irisbus (see Figure 3.33). The engine has a maximum power of 80 kW and is supplied by a 350V electric network.

Diesel-electric architecture is marketed by Irisbus in the Civis 18.5 m version comprising the optical guidance developed by Matra Transport (currently Siemens). These vehicles were sold in Clermont-Ferrand (with 225 kW diesel Euro3 engine of), in Las Vegas (10 units in mid-2003) and in demonstration in Rouen in 2004.

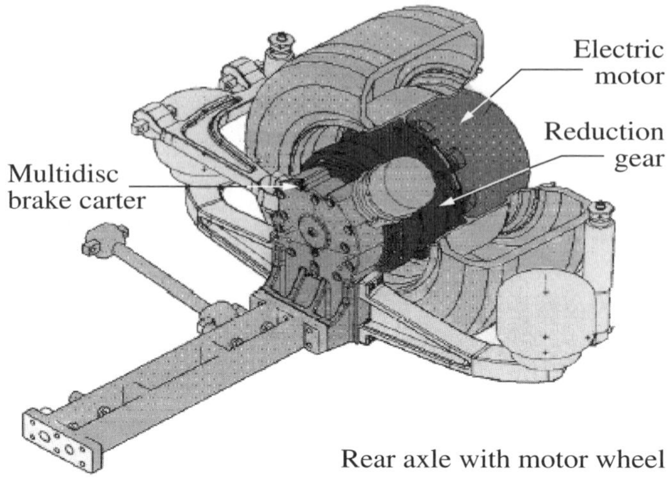

Figure 3.33. *Motor incorporated in the wheel for buses*

While the choice of the operation zones of the thermal engine makes it possible to reduce its consumption, the loss of efficiency on the electric drive makes it difficult to obtain consumption lower than that of the traditional vehicle. We can notice in addition that Irisbus markets vehicles in a trolley-bus version, using the same electric drive train. More than 60 vehicles (12 and 18 m) are currently used in Lyon under the name of "Cristalis".

Daimler-Benz

Mercedes is also developing a 9 m vehicle, the "CITO", which is propelled by a 125 kW thermal engine (see Figure 3.34). This vehicle was tested on-bench and in the agglomeration of Nimes within the framework of an ADEME project. The results showed a strong fall in the NOx emissions but no significant gain in consumption.

Figure 3.34. *Diesel-electric Mercedes CITO vehicle*

3.4.6.2.3. Series-hybrid implementation

Many prototypes have been produced and manufacturers are starting to propose hybrid vehicles to operators in their catalogue; we can mention:

– Irisbus, which, via its Altra subsidiary company, markets a 7.4 m hybrid bus (Europolis). This vehicle is driven by a 96 kW Siemens electric motor, a thermal (Euro3) 2.8-liter Sofim engine with a 30 kW electric generator and 85 Ah Exide lead-acid gel batteries;

– the same manufacturer, Altra, has developed a 12 m hybrid on the basis of an Iveco CityClass vehicle driven by a 164 kW electric motor, the same Sofim engine and 60 kWh of batteries;

– Ontario Bus Industries (acquired by DaimlerChrysler in 2000), who have been developing a 12 m hybrid bus since 1996, with a drive train provided by General Electric. This bus, now motorized by Lockheed-Martin (who have since become British Aerospace) has been the subject of a significant development on the network of the city of New York. More than 100 vehicles were ordered, with an objective to have nearly 400 vehicles in 2005; the project experienced a delay and in mid-2004 only 15 vehicles were in operation, the objectives for 2005 being deferred to 2006. This bus, in its 2001 version, is very pragmatic in design; it preserves the rear transmission axle (no low integral floor) and does not have all-electric mode (see Figure 3.35). Comparison tests with a diesel bus were carried out on roller test benches, which indicated that a gain in consumption varying from 20 to 40% following the cycles of use can be obtained, the slowest cycles being most favorable to the hybrid. The measured emissions are also lower for the hybrid. It will be noted that in the particular context of American buses moving in large, congested agglomerations with consumption varying from 65 to more than 160 l/100 km (average of about 85 l/100 km in New York), a gain in consumption is more easily obtained than in Europe with values of about 45 to 55 l/100 km.

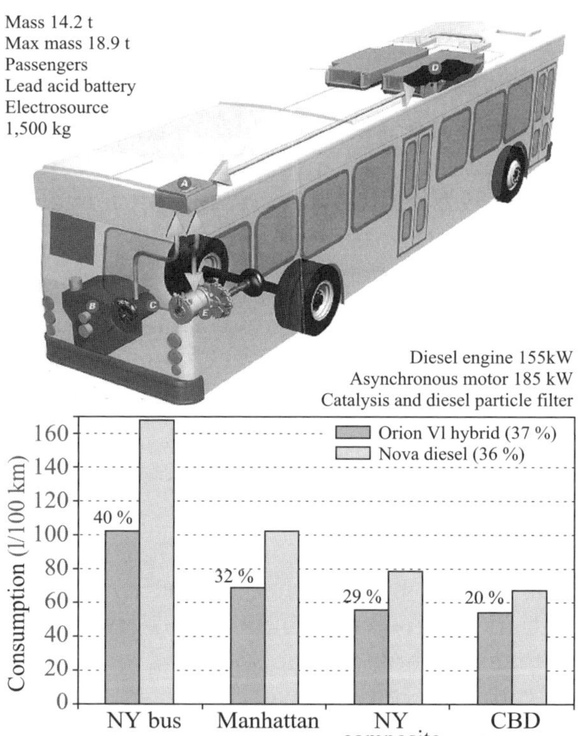

Figure 3.35. *New York hybrid bus*

General diagram comparison of consumption on cycle:

The main issues raised by operators are on the one hand the replacement of the battery, cost for lead-acid: 15,000 Euros in 2001 with a life expectancy of 1 to 2 years, and on the other hand a still significant failure rate for these innovating vehicles, (1.7 times more than a traditional bus).

These hybrid buses are in competition with natural gas (NG) buses; they are more expensive with an additional purchase cost of about 30% compared with 15% for the NG solution (price of diesel vehicle: 290,000 Euros in 2001). However, the hybrids have a very important advantage for operators located in agglomerations, which is not requiring an adaptation of the refueling station (safety of the buildings, space and time needed).

The firm NEOPLAN developed a 8 m (N6108) which was tested on-bench and in the agglomeration of Avignon within the framework of an ADEME project.

This vehicle is driven by an IVECO 2.8-liter 90 kW thermal engine, 20 kWh of HAWKER Génésis 70 Ah batteries and two asynchronous motors of 60 kW peak.

The tests carried out indicate that this vehicle presents a gain in NOx emissions, consumes electrical energy on the network during its battery recharge (74 kWh/100 km) but does not allow an overall energy gain.

The first series buses have thus reached the stage of *real-life* experiments. It will be advisable to follow their operations attentively in order to check if the advantages brought by this solution in terms of consumption, emissions, accessibility and energy management will balance the additional costs (purchase, operation) compared to the other solutions to be implemented to respect future standards (diesel post-treatment, sulfur-free fuel, alternative fuels , etc.).

3.4.6.2.4. Implementation of parallel and parallel-series hybrid

This type of architecture has the advantage of not modifying the configuration of the vehicle drive train in any depth.

Some achievements make it possible to be integrated instead of the traditional transmission, not requiring modifications from the manufacturer other than the integration of the battery pack.

This configuration does not make it possible, however, to have a low integral floor, all-electric mode with autonomy or an optimum energy recovery during braking.

It will be noted that the manufacturers introduced hereafter have developed architectures offering possibilities of parallel-series coupling, with two electric motors, which have the following advantages:

– good adaptation to urban or peri-urban uses by the use of electric mode or combined thermal-electric mode;

– good adaptation to road or motorway uses by the use of thermal mode with direct engine coupling to the transmission.

These configurations are extremely optimized from the energy and compactness point of view; they implement energy flow sharing systems (mechanical-electric) which are well controlled by transmission manufacturers. If the cost and reliability constraints on the system unit (transmission, battery, auxiliaries) are well controlled, such solutions can offer an interesting potential adaptable to multiple uses and a large variety of vehicles (buses, commercial vehicles, heavy trucks), with a minimum of modifications to the architecture.

Parallel hybrid architecture was less developed than the series configuration for urban transport; we can mention:

– the Japanese manufacturer Hino (Toyota group), who have been working since the beginning of the 1990s on a concept with an asynchronous electric motor linked to the thermal engine and a lead-acid battery (HIMR, Hybrid Inverter-controlled Motor & Retarder System concept). This transmission equips a 10.6 m urban bus with an electric motor of 30 kW peak and an 11.5 m peri-urban bus with a 22 kW engine. More than 300 vehicles have been delivered to urban operators and national parks; in 1999 the urban version was sold 44% more expensive than the diesel version (121,000 Euros). Hino more recently developed a parallel-series system implementing two electric motors;

– the transmission manufacturer Allison (General Motors group), who have developed an architecture of the parallel-series type including a thermal engine, two electric motors and an accumulator (see Figure 3.36).

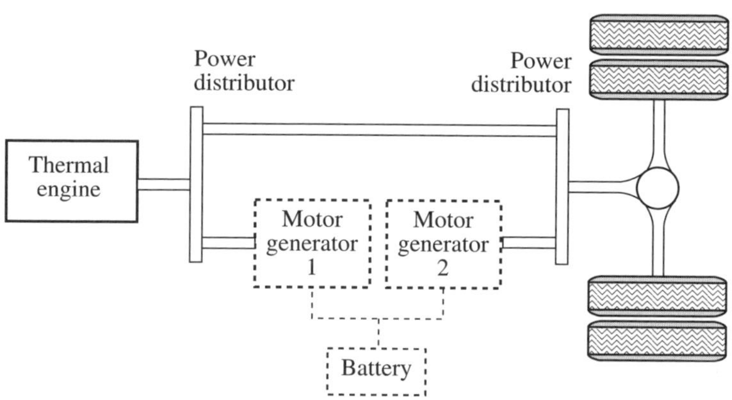

Figure 3.36. *General diagram of Allison parallel-series transmission*

Such a system makes it possible on the one hand to favor the use of the electric flow under urban operation, which makes it possible to have a continuous transmission, and on the other hand to privilege the use of a direct mechanical link, with a very good efficiency, in road or motorway use. The distribution of flows between the electric and mechanical way is carried out by distributor boxes (see Figure 3.36).

Allison has developed a version adapted to urban buses (EP40) with a 209 kW thermal engine, being able to develop up to 260 kW assisted by the electric system and a version adapted to the interurban bus (EP50) with 246 and 298 kW respectively. The tests carried out on this transmission equipping a NewFlyer 12 m bus showed a very strong reduction in the NOx emissions and particulates as well as a reduction of more than 30% in consumption (44.8 l/100 km compared with 67.2 for a diesel bus on a CBD14 cycle). An evaluation program, in collaboration with American operators, was planned for 2002 and production start-up in 2003.

3.5. Fuel-cell vehicles

3.5.1. *History, introduction*

Serious estimates lead us to think that the number of cars will grow from 500 million units currently to approximately a billion around the year 2030. This increase will not be without consequence on the environment, supposing that the type of motorization and the form of primary energy are not modified. Whilst it is obvious that a partial solution to this problem is the development of public transport, nobody envisages the end of the light vehicle and their number will continue to increase significantly.

Under these conditions, the quest for a lesser pollution becomes a question of an improvement in the efficiency of the on-board energy converter and the choice of a non-pollutant energy carrier compatible with current fossil fuel and the renewable energy sources of the future.

The ideal pair thus defined is the *fuel-cell* supplied with *hydrogen* gas.

Historically, the first vehicle equipped with a fuel cell was an Austin A40 modified by Karl Kordesch, in 1970, with an alkaline fuel-cell (see section 4.4.2) This vehicle was equipped with a 25 Nm^3 pressurized hydrogen tank installed on the roof and which ensured 300 km of autonomy; this prototype operated for three years without any particular problems.

In spite of these first encouraging results, the following prototypes were equipped with "acid polymer" or "PEM" cells (see section 4.4.3), whose development corresponds to the most important research tasks of the late 1980s. The most dynamic manufacturer by far in this development, in the early 1990s was, and still remains, Daimler-Benz, now DaimlerChrysler. Since then, the vast majority of the world's manufacturers have engaged on the same pathway.

3.5.2. *Choosing the kind of fuel cell*

The criteria that manufacturers had to take into account to choose the fuel cell most adapted to the light vehicle are numerous; the main ones are the following:

– technology for which the cost projections are lowest and in any event of the same order of magnitude as the current cost of the thermal engine;

– specific mass and volume equivalent to those of a thermal engine;

– low enough operating temperature to authorize an acceptable startup time (from 1 to 2 minutes) but high enough to allow a residual heat evacuation mode compatible with the volume available;

– technology compatible with correct operation in the traditional temperature range (- 30 to +40°C);

– operation with air as a combustive and insensitivity to gases and pollutants included in the air.

Taking these criteria into account led virtually all manufacturers to make the choice of the technology known as PEM (see section 4.2.4). The Canadian manufacturer, Ballard Power Systems, currently equips the majority of the prototypes comprising PEM systems. Nevertheless, whilst the fuel cell is the main component of this new technology, the fuel is at the center of the debate on this subject. If the final fuel is hydrogen, which all agree is the fuel of the future, this gas is not distributed today on the road. Consequently, two major approaches currently clash:

– whatever the primary fuel selected, hydrogen is manufactured in-station and distributed to users in a form that is to be specified (gas under pressure, liquid, absorbed);

– the vehicle carries a hydrocarbonated fuel (hydrocarbon or alcohol) and transforms it on-board (reforming) into hydrogen, as and when it is needed.

According to the option selected, the design of the vehicle will not be the same and according to the financial resources mobilized by one manufacturer or another, they test the two approaches or choose one only. Today, at the beginning of the 21st century, it is far from simple to predict which of the two methods will prevail between now and the final dawning of hydrogen, expected to occur just before the middle of the century.

3.5.2.1. *The principle of a fuel-cell vehicle*

Just like in a thermal motor vehicle, the useful energy is the result of the reaction of on-board fuel with the oxygen in the air. But instead of an explosive reaction at the top of the piston, it is a controlled reaction within an electrochemical converter (the fuel cell, see section 4.4.3) which is the energy source of the vehicle. This energy in electric form supplies an electric motor which moves the wheels.

In a more complex but more powerful alternative, an electric storage system (battery or super capacity) is buffered. The vehicle is known as hybrid and it then makes it possible to satisfy various complementary functions:

– another optimization of the fuel-cell operation, and possibly another dimensioning insofar as this battery can bring an occasional supplement of power that the fuel cell will not have to provide, thus making it possible to design it with a lower power;

– a startup assistance, in particular to await the temperature setting, either of the fuel cell in cold weather, or of the reformer if the vehicle is equipped with it;

– a recovery of energy during braking, given that the rate of recovery in urban circulation can reach more than 20% of the power consumption.

A light fuel-cell vehicle (passenger car) consumes approximately 1 kg of hydrogen per 100 km.

3.5.2.2. *The major car manufacturers' fuel-cell programs*

General Motors

General Motors started its fuel-cell vehicle program in 1990 with the assistance of the Department of Energy (DoE), within the framework of the PNGV (Partnership for a New Generation of Vehicles) program, in collaboration with Delphi Automotive Systems.

The principal subcontractors were DuPont de Nemours and Ballard Power Systems.

The first phase was completed in 1994 by the demonstration of feasibility of a 10 kW fuel-cell prototype. Phase II (1995-1997), worth $36 million, led to the demonstration of a 60 kW prototype system (two 30 kW modules with all its auxiliaries) hydrogen-fuelled by a methanol reformer provided by Arco Products Co (subsidiary company of Atlantic Richfield and Exxon).

Phase III (1997-1999) made it possible to produce several prototypes of fuel-cell vehicles at the beginning of 1998:

– a *Chevy Van* equipped with a methanol reformer;

– an *EV-1* vehicle (battery-powered electric vehicle) modified with a fuel cell and a methanol reformer;

– a first prototype based on an Opel *Zafira* (presented at the Paris Motor Show in September 1998), equipped with two 25 kW Ballard fuel-cells and a 6.3 kWh Ovonic NiMH battery;

– the *Precept* prototype, presented in January 2000 at Detroit, equipped with a 75 kW "house" fuel-cell (made up of 400 cells), fuelled by a hydrogen tank stored on hydrides (4.9 kg stored). The vehicle is driven by a 85 kW electric motor;

– a second *Zafira* prototype presented at the Geneva Motor Show (March 2000), called *HydroGen1*, fuelled with liquid hydrogen (5kg stored) and equipped with a 86 kW "house" fuel-cell whose weight and volume characteristics are 59 liters and 86 kg respectively. The manufacturer ensures that the full power is obtained at the end of 30 seconds after a startup with -20°C. Note that the vehicle thus equipped weighs only 150 kg more than the basic model;

– General Motors is currently working on a prototype built based on the Chevrolet S-10 Pick-up, which should begin its tests at the beginning of 2002, and which is to be equipped with a reformer with gasoline whose startup time is to be lower than 3 minutes;

– a improved version of the Hydrogen1 prototype, baptized Hydrogen3, was produced from 2002, in two versions, one with liquid hydrogen and the other with 350 bars pressure hydrogen;

– in 2002, General Motors forwarded the interesting concept, baptized Skateboard, of a frame in which all the functions were integrated, and that one equips with the desired body: two versions were thus forwarded: Hy-Wire in 2002 and the "Sequel" at the beginning of 2005, both with respectively 350 and 700 bars pressure hydrogen under.

The center for General Motors research on fuel-cell vehicles is located in Germany at Russelsheim (GM Fuel Concealment Initiatives) an agreement was signed in June 1998 between Toyota and General Motors for the joint development of fuel-cell vehicles.

Comments:

– This set of prototypes shows that General Motors initially accumulated experience based on Ballard technology to then develop a fuel-cell technology in their own laboratories.

– In complement, General Motors developed a reformer technology (with Arco and Exxon, initially with methanol then with gasoline) and two hydrogen storage technologies (on hydrides and in cryogenic form), applied to hybrid concepts (with a NiMH battery) or not.

– In short, General Motors tested almost all of the possible fuel-cell vehicle alternatives and thus accumulated great experience on the subject.

However, it does not seem to have deduced from this a choice of any sort, whilst nevertheless showing – as all the American manufacturers – that gasoline reforming will be of a certain interest, should its feasibility suddenly be shown.

Ford Motor Co.

Ford Motor Co. started its work in 1990, under the aegis of DoE like General Motors and Chrysler, with a very similar program.

In the first phase, Ford compared technologies of five suppliers: International Fuel Cells (IFC), Mechanical Tech. Inc. (MTI), Energy Partners, H Power and Tecogen on the basis of a 12 kW fuel-cell module, at the conclusion of which two were retained: Plug Power/MTI and IFC.

In the second phase, which began in 1995, these two suppliers each provided a 50 kW system under 250V, the IFC system being linked to a pressurized hydrogen tank.

At the same time, in December 1997, Ford announced its alliance with Ballard and Daimler with the new company, DBB Fuel Cell Engines[4], so that all the vehicle prototypes presented thereafter were equipped with Ballard fuel cells:

4 This corporate body became XCELLSiS, it was taken over and integrated in 2001 by Ballard.

– the first prototype P 2000 was presented in January 1999 at the Detroit Motor Show. The fuel cell was supplied by a small, pressurized hydrogen tank (1.4 kg of H_2 on-board) and was made up of three 25 kW modules;

– the second prototype, baptized *TH!NK FC5*, based on a Ford *Focus*, was presented in January 2000 at the Detroit Motor Show. Its 75 kW fuel cell was supplied by a methanol reformer;

– the third prototype, baptized *TH!NK FCV*, was presented in February 2001 with a small, pressurized hydrogen tank using Dynetek technology, which authorized only 150 km of autonomy;

– in 2002, Ford brought out the Focus FCV series, of which about 15 units were equipped with a Ballard fuel cell and hydrogen tanks under 350 bars;

– in 2004, about 30 other Focus FCV were built for various tests in real situations.

Ford Motor Co. is working, along with Mobil Corp., on the development of a gasoline reformer.

DaimlerChrysler AG

Daimler and Chrysler each separately embarked, from 1993, on work on the fuel cell (Daimler on clean appropriations in complement to the signing of an agreement with Ballard and Chrysler with the assistance of the US/DoE on an MTI/Plug Power fuel-cell technology).

During this time, no prototype was presented by Chrysler and three vehicle prototypes were presented by Daimler:

– the prototype *Necar 1*, presented in March 1994, based on a Mercedes-Benz MB 180 Van vehicle. This vehicle was essentially a moving experimental base, without any concern for integration, equipped with 12 Ballard modules for a total available of 50 kW, supplied by a hydrogen tank under a pressure of 250 bars;

– the prototype *Necar 2*, presented in May 1996, based on a Mercedes-Benz V-Class vehicle, on which serious integration work had been carried out. It was equipped with two 25 kW modules supplied by a pressurized hydrogen tank (approximately 3 kg stored);

– the prototype *Necar 3*, presented at the Frankfurt Motor Show in October 1997, based on the Mercedes-Benz A-Class vehicle. It was equipped with a 50 kW fuel cell placed in the chassis frame at the center of the vehicle, supplied with hydrogen by a methanol reformer placed at the rear of the vehicle. The 40-liter tank gave it an autonomy of 400 km.

The repurchase of Chrysler by the Daimler group and the birth of DaimlerChrysler AG at the beginning of 1998, allowed the regrouping of work on fuel-cell vehicles and other prototypes were presented:

– the prototype *Necar 4,* presented in March 1999, in Washington, based on the same A-Class vehicle, which differed from the previous one by a more advanced fuel cell (70 kW), supplied this time by a tank containing 7 kg of liquid hydrogen. The manufacturer points out (1) that the diesel version of this A-Class consumes 5.6 liters per 100 km, whereas the Necar 4 consumes the energy equivalent of 3.2 liters of fuel, and (2) that the total efficiency (WtW) is 37.7% on the basis of the European standard cycle (NEDC[5]), compared with 26% for a current powerful diesel vehicle and 22% for a gasoline vehicle. This Necar 4 vehicle weighs 460 kg more than the initial A-Class version, however;

– the prototype *Necar 4 Advanced*, presented in November 2000. This vehicle is distinguished from the previous one by a hydrogen tank under a pressure of 350 bars; indeed, the prototype Necar 4 had been primarily developed to be set up in the California Fuel-Cell Partnership (CaFCP), a structure for the demonstration of fuel-cell vehicles (light and heavy vehicles) located in California. For financial reasons, however, the CaFCP could not install a liquid hydrogen distribution system, but only a pressurized hydrogen distributor, hence the modification carried out on the Necar 4. It should be noted that DaimlerChrysler plans to increase the storage to 700 bars, as on the French PSA prototype;

– the prototype *Jeep Commander 2*, presented in November 2000, equipped with a methanol reformer and new Ballard technology Mark 900 fuel-cell technology;

5 NEDC : New European Driving Cycle.

– the prototype *Necar 5*, presented in November 2000, with characteristics very close to the model *Necar 4*, but equipped with more compact Ballard Mark 900 technology;

– a small prototype using a new PEM/DMFC (direct methanol) fuel-cell technology presented in November 2000, with a power of 3 kW.

Comments:

1. As a result of this work, DaimlerChrysler proceeded with a restricted number of tests on the various possible solutions, contrary to General Motors, for example, and – *a priori* – made the following choices at the beginning of 2005:

– the on-board methanol, whether it is used in a reformer or directly (DMFC);

– the hybrid vehicle with Ni-MH battery and 350 hydrogen fuel tank.

This choice preserves the future by developing the technology of hydrogen storage under high pressure (350 to 700 bars).

2. DaimlerChrysler developed, in parallel, a fuel-cell bus network initiated by the Nebus prototype, presented in July 1997, whose spectacular outcome is the signature, in March 2001, of the European FP5 NEFLEET contracts: introduction of about 30 Citaro-type fuel-cell buses in 10

European cities (with France not taking part) by 2004. These buses are equipped with a 250 kW fuel cell.

Volkswagen

This manufacturer apparently began work on the subject around 1995 with the European CAPRI contract for an electric VW Golf vehicle equipped with a 15 kW Ballard fuel cell used as a range extender and fuelled by a Johnson Matthey "Hot Spot"-type reformer.

In November 2000, Volkswagen presented its first prototype based on the Bora model, baptized *Bora HyMotion*.

The 30 kW fuel cell was provided by IFC (International Fuel Cell) and it is a hybrid with a 3 kWh NiMH battery; it is fuelled from a liquid hydrogen tank. Its asynchronous electric motor has a power of 75 kW.

This first prototype was followed by the prototype Bora HyPower, in 2001, equipped with a 40 kw fuel cell hybridized with a battery and a set of supercapacities, followed by the Touran HyMotion prototype, in 2004, supplied with a 350 bars hydrogen and hybridize with a Ni-MH battery.

Renault

Renault began its work on the fuel cell in 1992 within the framework of the national VPE[6] program, in collaboration with PSA, the CEA and CNRS.

In 1994, Renault joined forces with several partners[7], within the framework of a European project baptized *Fever*, to study and produce a prototype of the fuel-cell vehicle based on the *Laguna* model.

---

6 VPE: French national project on green vehicles.

7 Project partners are Air Liquide, Ecole des Mines, DeNora (now Nureva), Ansaldo and Volvo.

Equipped with a 30 kW DeNora fuel cell with a NiMH battery and supplied with a cryogenic hydrogen tank (8 kg of hydrogen stored), the prototype was tested from mid-1998.

This vehicle was treated like the Daimler Necar 1, i.e. like a moving platform, without optimization for component integration.

After the purchase of the Japanese manufacturer Nissan in 2000, the two manufacturers amalgamated their R&D work on fuel-cell vehicles.

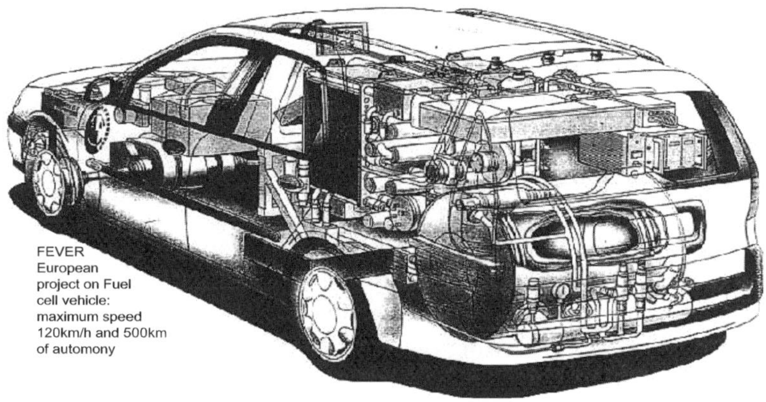

PSA Peugeot Citroën

As is specified in the preceding section, PSA Peugeot Citroën began its work on the fuel cell in 1992, in close cooperation with the manufacturer Renault.

HYDRO-GEN project

In 1996, PSA joined forces with several partners[8], within the Hydro-Gen (European framework project), to study and produce a prototype fuel-cell vehicle on the basis of the

---

8 Project partners are: Air Liquide, DeNora (now Nureva), Ansaldo (1st phase), Solvay, Renault, Sorapec, CEA and CNRS.

Peugeot Partner model, but on an integrated concept, compared with the prototype of the Fever program.

On this prototype, a 30 kW DeNora fuel-cell is supplied by a pressurized hydrogen tank under 700 bars in four cylindro-spherical composite-structure tanks, located in the chassis frame of the vehicle.

Integration made it possible to leave useful space inside the vehicle, very close to the basic model. This prototype began its tests in May 2001.

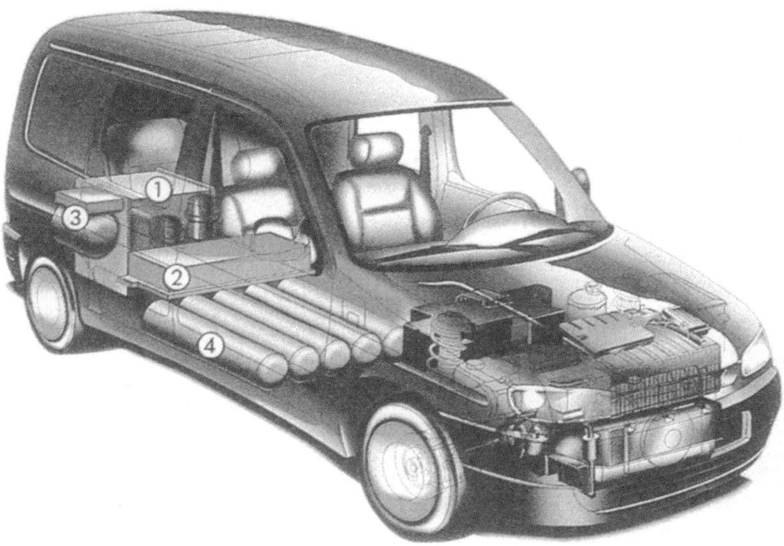

Performance of the Hydro-Gen project fuel-cell is shown in the table below.

Weight	130 kg
Net Power	30 kW at 70 Volts
Stack efficiency	56%
Plate load per electrode **Total**	0.3 mg/cm^2 45g

TAXI PAC & H20

These technological demonstrators use the Fuel Cell as a Range Extender or Auxiliary Power Unit (APU).

The fuel cell recharges the batteries, which restore this energy to the electric traction motor according to the need. Vehicles equipped with a fuel-cell range extender are very well adapted to the urban environment.

The use of an electric motor enables them to be silent, which is an important quality for a vehicle driven in town.

Their autonomy, about 300 km, is sufficient for urban and peri-urban driving.

Lastly, the engine consumes energy only when the car is moving, and it makes it possible, when the car is in deceleration phase, to recover energy.

The PSA PEUGEOT CITROËN group produced a demonstrator based on an electric Peugeot Partner.

Something of a “London taxi for the 21st century”, it preserves its original electric motor (with a nominal output of 22 kW) and receives a low-size fuel cell (5.5 kW), nickel-metal hydride batteries, and a removable “rack” made up of hydrogen bottles.

Hydrogen is thus stored on-board in the form of gas compressed at 300 bars. The exchange of an empty rack for a full rack is done very quickly. The filling of the hydrogen bottles is thus carried-out outside of the vehicle without any major time constraints.

The second demonstrator, H_2O, is a firefighter response vehicle. It uses the same drive-train system as the TAXI PAC but hydrogen is produced on-board from a solution of sodium boro-hydride.

PEUGEOT H_2O

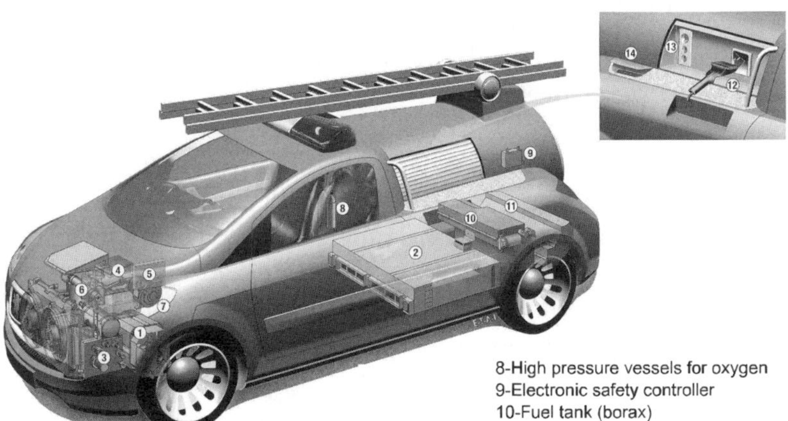

1-Electric motor
2-Drive battery
3-Fuel cell system (range extender + APU)
4-Electric vehicle controller
5-Hydrogen system controller
6-Hydrogen supply unit
7-Air pump

8-High pressure vessels for oxygen
9-Electronic safety controller
10-Fuel tank (borax)
11-Spent tank
12-Charging sockets
13-Electrical connections for auxiliares
14-Fuel filler flap

Toyota

Japanese manufacturer Toyota presented its first vehicle prototype in November 1996, at the 13th International Electric Vehicle Symposium in Osaka.

It used a basis of the RAV-4 LV vehicle and comprised a 20 kW “house” fuel-cell hybrid with a 110 kg lead-acid battery.

The fuel cell was fuelled by a hydrogen tank (2 kg stored for an autonomy of about 200 km) absorbed on hydrides. Other prototypes followed:

– a second prototype with the same RAV-4 basis, presented in October 1997, and equipped with a 25 kW fuel cell, fuelled from a methanol reformer (of the vapor-reforming type) and hybrid with a NiMH battery. The electric motor (synchronous, with a permanent magnet) had a power of 50 kW. The quantity of methanol stored allowed an autonomy of 500 km;

– a third prototype, baptized FCHV-3 and derived from the Prius, was presented in March 2001, at the Tokyo Electric Vehicle Symposium. It is equipped with a 90 kW “house” fuel-cell, hybrid with an NiMH battery and fuelled with hydrogen from a storage on hydrides the storage performance in mass is 2%. In the same year, 2001, were forwarded two versions of: the FCHV-4 one equipped with 250 bars hydrogen tank , and the other with a gasoline without sulfur reformer.

– in November 2002, Toyota started a phase of hiring about 20 specimens of its FCHV-4, in the United States and Japan. It model is equipped with 350 bars hydrogen tank, “in house” fuel cell of 80 kw and hybridize with a NiMH battery;

– in November 2003, Toyota presented the *Fine-S* prototype of a rather similar concept to Hy-Wire from General Motors. This prototype was followed up, in October 2003, by *Fin-N*, equipped with a 700 bar hydrogen tank.

In addition, Toyota developed a series of fuel cell buses, in service in Tokyo since the beginning of 2003, baptized FCHV-BUS.

Toyota developed its fuel-cell technology in collaboration with Denso Corp. and Aisin Seiki Co. and signed a collaboration agreement on fuel-cell vehicles with General Motors in 1998.

Comment:

Toyota is the first Japanese manufacturer to have launched the production of fuel-cell prototype vehicles. It joined the majority of the manufacturers which retained since 2004:

– hybrid vehicle with a battery;

– 350 bars hydrogen storage with 700 bars evolution.

Mazda

This manufacturer presented its first prototype in December 1997, at Eco Japan '97. Baptized *Demio FCEV*, it was equipped with a 20 kW fuel cell (in four 5 kW modules), hybrid with a super-capacity of 20 kW, the unit supplying a 40 kW electric motor.

The fuel cell was fuelled by a hydride hydrogen tank (eight modules) which gave it a range of 170 km. Other prototypes followed:

– a second prototype of the *Demio*, presented at the Tokyo Motor Show, in October 1999, equipped with a 50 kW fuel cell;

– a third prototype baptized *Premacy FC-EV*, presented at the beginning of 2001, and resulting from a collaboration with the Ford TH!INK group. It is equipped with a Ballard fuel-cell supplied by a methanol reformer and which provides energy to a 65 kW induction alternate-current electric motor. It should be noted that this vehicle is authorized to run on public roads, which does not seem to be the case, in 2001, of the competitor prototypes. In 2005, no other prototype of fuel-cell vehicle was shown. One can on the other hand announce the interest of Mazda for hydrogen with the continuation of the development of his model RX-8 driven by a rotary positive ignition engine, supplied with hydrogen.

Honda

This manufacturer launched into the development of fuel-cell vehicles later than the two preceding Japanese manufacturers.

The first official marks of interest of this manufacturer date back only to July 1999, but in a short space of time, three prototypes were presented:

– the prototype baptized *FCX-V1*, presented in September 1999, equipped with a 60 kW Ballard fuel cell, fuelled by a hydride hydrogen tank;

– the prototype baptized *FCX-V2*, presented at the same time, equipped with a 60 kW "house" fuel cell, fuelled by a methanol reformer (autothermal type);

– the prototype baptized *FCX-V3*, presented in February 2001, equipped with a 70 kW "house" fuel-cell, supplied with a 250-bar pressurized hydrogen tank and hybrid with a super caps;

– the prototype baptized FCX-V4, was forwarded in October 2001; it is about a version improved of precedent with a 350 bars hydrogen tank;

– from the end of 2002, Honda manufactured several dozen vehicles baptized Honda FCX from a yet further improved version of FCX-V4. It is equipped with an "in house" fuel cell of 86 kw, of which, in February 2005, 14 unit were in the United States, in hiring among various customers.

Comment:

This program shows that, in a first phase led at a quick pace, Honda acquired experience and tried to make up for lost time by testing all the possible solutions: Ballard reference and "house" fuel cell, hybrid or not with an electrochemical storage, fuelled with several types of hydrogen supply (reformer, hydride hydrogen and pressurized hydrogen).

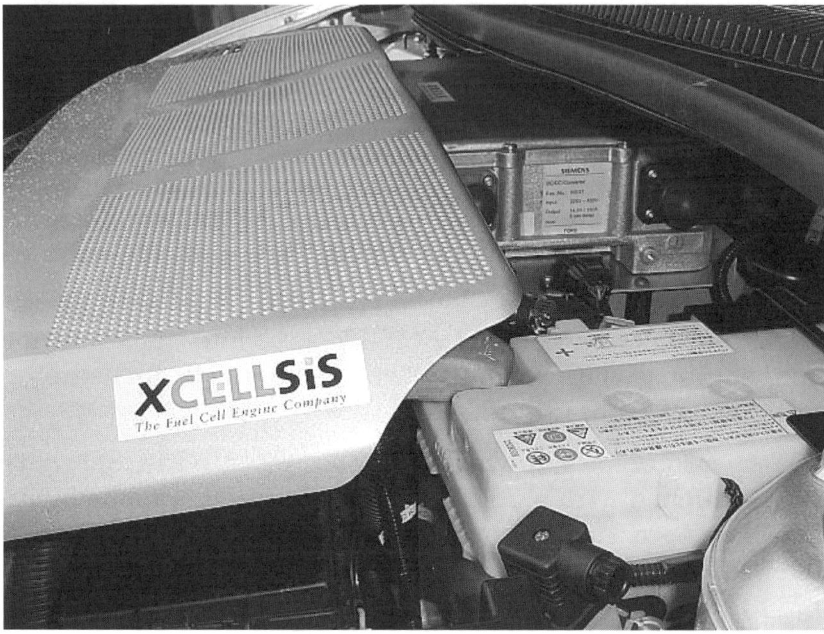

Nissan

The first official signs of Nissan's interest in fuel-cell vehicles go back to the ordering of fuel-cell modules from Ballard in March 1997. Three prototypes have been presented since then:

– in May 1999, a prototype built based on an *R'Nessa/Altra EV*, equipped with a 10 kW fuel cell, hybrid with a lithium-ion battery and fuelled by a methanol reformer;

– in 2000, a second prototype baptized *Xterra* was presented, the fuel cell being fuelled by a pressurized hydrogen tank

– in 2002 a third prototype was presented, baptized XTrail FCV, supplied with hydrogen under 350 bars and hybridize with a Li-ion battery, whose some specimens were put in hiring in 2004. This vehicle has the characteristic of being able to be equipped with three models of fuel cells: Ballard, UTC Fuel "in house".

It should be noted that Renault and Nissan announced in November 2000, the launching of a common fuel-cell vehicle development program, with a budget of 800 million Euros until 2005.

Mitsubishi

Mitsubishi Motors joined Mitsubishi Heavy Industries Ltd at the end of 1998 to develop a fuel-cell vehicle.

Mitsubishi Electric (for the electric motor) and Nippon Mitsubishi Oil (for the development of a reformer, probably with methanol) also contributed to this project. In October 1999, a first prototype was presented at the Tokyo Motor Show.

Baptized MFCV (for Mitsubishi Fuel-Cell Vehicle), it was equipped with a 40 kW fuel-cell hybrid with a lithium-ion battery and fuelled by a methanol reformer. Mitsubishi Motors Corp. presented in 2003, the prototype baptized Grown Minivan, equipped with a Ballard fuel cell hybridized with a NiMH battery and supplied with a 350 bars hydrogen tank. This prototype received, at the end of 2003, certification to circulate on the Japanese roads.

It should be noted that in May 2000, DaimlerChrysler and Mitsubishi Oil Co. signed an agreement for the development of fuel-cell vehicles.

Daihatsu

Daihatsu presented a fuel-cell prototype vehicle (micro-van), baptized *MOVE*, in October 1999.

It was equipped with a "house" fuel-cell hybrid with a 16 kW NiMH Panasonic battery and fuelled by a methanol reformer (vapor-reforming technology). In November 2001, Daihatsu presented its second prototype baptized Move FCV-K-II, equipped with a Toyota 30 kw fuel cell hybridize with a NiMH battery and supplied by 250 bars hydrogen tank.

It should be noted that Daihatsu signed a collaboration agreement with Toyota on the subject.

Hyundai Motor Company

Korean manufacturers[9], supported by the public authorities, seem to invest massively in the development of fuel-cell vehicles[10].

The first to have presented a prototype is Hyundai Motor Company.

---

9 Hyundai Motor company and Kia Motors subsidiary with the aid of Daewoo motor and petrolum, Korean companies (SK Corp and LG 6caltex Oil).

10 Korean Herald from February 7th 2000.

This prototype, baptized *Hyundai Santa Fe Fuel-Cell Vehicle*, was presented in March 2001.

It is the fruit of a collaboration with Enova Systems of Torrance, International Fuel-Cell and IMPCO Technologies Inc.

The fuel cell is fuelled by a small, pressurized hydrogen tank (autonomy 160 km). It is intended to remain in demonstration in the USA within the framework of the California Fuel-Cell Partnership.

In April 2004, Hyundai presented its prototype 4x4 baptized *Tucson FCEV*, equipped with UTC Fuel-Cells, hybrid with a Li-ion battery and fuelled by a 350-bar hydrogen tank.

The construction of several dozens of specimens was planned for 2005.

The Chinese program

As from 2003, China has been active in the field of fuel-cell vehicles, whether with two-wheeled vehicles, light vehicles, service vehicles or buses.

Their development program is very ambitious.

Research centers (Tsing Hua and Tongji Universities) and manufacturers (Shanghai Automotive Industry Corp., Shen-Li Company Ltd., FAW Group Corp.) collaborate and many concept cars are regularly presented (*Start* 1 and 2, *SLFC-Tourist* 1 and 2, in 2003 and 2004, fuelled with pressurized hydrogen).

It should be noted that a collaboration agreement relating to the development of fuel-cell vehicles was signed between the Chinese government and the Japanese company, Toyota.

3.5.2.3. *Overall assessments*

Table 3.9 summarizes the main features of the various concept cars of PEM fuel-cell vehicles known in February 2005.

Of all 21 listed manufacturers, from eight countries (United States, Japan, Korea, China, Russia, Germany, France and Italy) and which have produced more than 300 light vehicles (on the basis of 80 different models) in around ten years (1994-2004), we can make, in 2005, the following observations:

– after the tests of all the possible modes of hydrogen production or storage on-board (gasoline or methanol reforming, decomposition of sodium borohydrides, storage on metal hydrides, storage in the form of liquid hydrogen, storage in the form of compressed gas), it is the method of hydrogen compressed with 350 bars in composite-frame bottles which is favored by virtually all the manufacturers, with a probable change to 700 bars in the next few years;

– the PEM fuel cells are provided primarily by the following manufacturers: Ballard, UTC Fuel Cells, Nuvera Fuel Cells, General Motors, Toyota and Honda;

– the fuel cells are practically all hybrid with batteries (NiMH and Li-ion), however, some are so with super-capacities;

– in 2004, six manufacturers started the phase of mini-productions of 30 to 60 identical specimens, rented to various customers: Hyundai, DaimlerChrysler, General Motors, Ford, Toyota and Honda.

In spite of this fast progress, the majority of vehicle manufacturers do not envisage real marketing before the 2020s.

3.6. Bibliography

NICOLON, Alexandre, *Le Véhicule électrique: Mythe ou Réalité?*, Paris, Editions de la Maison des Sciences de l'Homme, 1984, 133 pages.

WOLF, Roland, *Le Véhicule Electrique gagne le Cœur de la Ville*, French Center of Electricity, 343 pages, 1995.

Michelin – Challenge Bibendum, OUI aux Voitures Propres, Editions Hoëbeke, 147 pages, 2002.

Le Mémento de l'Hydrogène, ten detailed technical data sheets (sheets 9.1.1 to 9.1.10) on vehicle manufacturers' "Fuel-cell vehicle" programs, available on the French Hydrogen Association's website: www.afh2.org.

La Vie Automobile, no. 150, August 13, 1906, page 524.

3.7. Summary table of fuel-cell (PEM) vehicle prototypes (as of February 2005)

Manufacturer	Model	Year	Hybrid-ization	Reform ing	Hydrogen storage
General Motors	Chevy Van				
	FCEV-1	1998	?	yes	no
	Zafira	1998	?	yes	no
	Precept	1998	NiMH	yes	no
	HydroGen1	2000	yes	no	hydrids
	S10-Pick up	2000	yes	no	cryogenic
	HydroGen3-	2001	no	yes	no
	cryo	2001	no	no	cryogenic
	Autonomy	2002	?	no	350 b
	Hy-wire	2002	no	no	350 b
	HydroGen3-	2002	?	no	pressure
	pres	2005	Li-ion	no	700 b
	Sequel				
Ford Motor Co.	P 2000 HFC	1999	no	no	pressure
	Th!nk FC5	2000	no	yes	no
	Th!nk FCV	2000	no	no	350 b
	FCV Hybrid	2002	NiMH	no	350 b
	Model U	2003	?	?	?

Anuvu Inc.	Pick-up truck	2003	?	?	?
Daimler Chrysler	Necar 1 Necar 2 Necar 3 Necar 4 Necar 4/Adv. Necar 5 Jeep Comm. 2 Sprinter Van Natrium F-Cell Jeep Treo	1994 1996 1997 1999 2000 2000 2000 2001 2001 2002 2003	no no no no no no NiMH Li-ion NiMH ?	no no yes no no yes yes no no no	pressure pressure no cryogenic pressure no no $NaBH_4$ 350 b ?
Audi	$A2H_2$	2004	NiMH	no	350 b
Volkswagen	Bora HyMotion Bora HyPower TouranHyM otion	2000 2001 2004	NiMH super cap NiMH	no no no	cryogenic 350 b 350 b
Fiat	Elettra H2FC Panda Hydrogen Seicento	2001 2003 2003	yes ? ?	no no no	pressure 350 b 350 b
Michelin-PSI	Hy-Light	2004	super cap.	no	pressure
Renault	Fever	1998	NiMH	no	cryogenic
PSA	Hydro-Gen Taxi H2O	2001 2001 2002	no NiMH NiMH	no no no	pressure 300 b pressure
Toyota	RAV-4 RAV-4 FCHV-3 FCHV-4 FCHV-5 FINE-S FINE-N MTRC	1996 1997 2001 2001 2001 2003 2003 2004	Pb NiMH NiMH NiMH NiMH ? Li-ion ?	no yes no no gasoline no ?	pressure no hydrides 350 b no 350 b 700 b ?

Mazda	Demio FCEV Demio FCEV Premacy FCEV	1997 1999 2001	super cap super cap ?	no no methan ol	hydrids hydrids no
Honda	FCX-V1 FCX-V2 FCX-V3 FCX-V4 Honda FCX Kiwami FCX 2005	1999 1999 2000 2001 2002 2003 2004	? ? super cap super cap super cap super cap	no methan ol no no no no	hydrides no pressure 250 b pressure 350 b pressure 350 b pressure pressure 350 b
Nissan	R'Nessa Xterra X-Trail	1999 2000 2002	Li-ion yes Li-ion	yes no no	no pressure 250 b 350 b
Mitsubishi	MFCV Space Liner Grandis minivan	1999 2001 2003	Li-ion non NiMH	yes methan ol no	no no 350 b
Daihatsu	Move Move FCV-K-II	1999 2001	NiMH NiMH	yes no	no pressure
Daewoo	DFCV-1	2001	?	?	?
Hyundai	Santa Fé Tucson 4x4	2000 2004	yes Li-ion	no no	350 b 350 b
Suzuki	Mobile Terrace MR Wagon FCV	2003 2004	yes	no no	pressure 700 b
Avtovaz Lada	Antel-1 Antel-2 (AFCV)	2001 2003	? NiMH	? no	? 400 b

Reva Electric Car (India)	Reva	2004	yes	no	pressure
Shanghai Shenli Tongji Univ.	Start 1 Start2 SLFC-Tourist 1 SLFC-Tourist 2	2003 2004 2004 2004	? Li-ion no no	no no	 pressure pressure pressure

Chapter 4

The Components of Electric-Powered Vehicles

Electric motors

Advantage of electric motors in road-vehicle drive-train systems

Before we begin presenting a description of electric motors, electronic converters and batteries used on-board vehicles, we will firstly give some elements that help appreciate the advantage of using electric traction in drive-train systems.

Let us note first of all that the mechanical torque of the electric motor results from the action of a magnetic induction flux on an electric current. In an electric motor, the flux imposes the dimensions of the magnetic material (iron); the current imposes the diameter of the wire (generally made of copper) which constitute the coil-windings.

Chapter written by Joseph BERETTA, Jean BONAL and Thierry ALLEAU.

Thus, electric-motor dimensions depend largely on the torque characteristics which we want to obtain.

The motor power is equal to the product of the torque by the rotation speed, such that at a given power, the size of the motor is as small as its rotation speed is high.

In a DC motor the rotational speed is limited:

– mechanically by removing risks on binder windings and commutated segments;

– electrically by the commutation of the current between the commutator segments.

In practice the electric motors used in road-vehicle drive-train systems have unit powers lower than 50-60 kW and their supply voltage most of the time remains lower than 200 Vdc.

Under these conditions it is possible to manufacture engines with a DC motor, which turn at 5,000 rpm.

Motors with AC current, because of the absence of a collector, can reach rotation speeds higher than those of the DC motor: 10,000 rpm and a perfectly feasible speed for the powers concerned in conventional road vehicles.

As such, the specific power of an AC motor is higher than that of a DC motor.

Whatever its nature – DC motor or AC motor – the electric motor presents a certain number of advantages:

It can provide a torque to all speeds and in particular when stationary. This property permits the elimination of the clutch in the torque transmission chain with the help of a suitable supply control voltage for a DC motor, or with help of the inverter in the case of an AC motor.

It can support short loads, and provide significant overtorques, about 2 to 4 times' the nominal torque, during the acceleration or starting period.

It presents a high thermal inertia; the maximum power is higher than the nominal continuous output.

The continuous power/maximum power depends on the size of the motor and its cooling mode. In general, for motors used in electric traction, this ratio is about 1.3.

4.1. Electric motors

The introduction in series of vehicles equipped with alternator-starters, electric-hybrid or quite simply electric drive-trains is currently very dependent on the evolution of electric parts, and in particular of electric motors.

Indeed, the technologies currently available for electric motors and power electronics induce additional vehicle costs that are incompatible with the costs/services accepted by the market.

These costs strongly limit a significant and durable marketing of this type of vehicle.

It is thus essential to develop technologies generating a significant reduction in the cost of electro-technical parts for automotive applications. These technologies must also be applicable transversely to various applications in order to generate a sufficient volume effect.

It is also essential to limit the over cost of auxiliary parts specific to these drive-trains (air-conditioning, braking, etc.) through appropriate architecture choices.

Developed technologies must lead to minimum cost and be applicable transversely for the applications concerned.

Let us recall the main constraints on electro technical parts for each application:

	Main constraints
Alterno-starter	– high compact – very wide speed operating range – mechanical integration: minimum gap to be guaranteed between stator an rotor
Hybrid	– good compactness – good efficiency
Electric	– very good efficiency

Table 4.1. *Current electric traction-motor technologies*

The DC motor was traditionally simplest to use in drive systems. With progress of electronics, various electric motors are now preferred in automotive drive: permanent-magnet or wound-rotor synchronous motors, as well as asynchronous or switch reluctance motors.

The synchronous permanent-magnet motor is penalized by a temperature limit of the magnets (135°C) and their cost. Moreover, its best output areas (see curve below), well to the pure electric vehicle, seem less well-adapted to hybrid or alterno-starter vehicles.

Figure 4.1. *Leroy-Somer synchronous motor*

The synchronous wound-rotor motor requires a more complex industrialization (winding of the rotor, rings and brushes).

The asynchronous motor is more difficult to operate at constant power on a wide speed range taking into account the defluxing conditions.

Without brushes, with a structure minimizing the short-circuit, the switch-reluctance motor presents less heating under operation compared to other motor technologies.

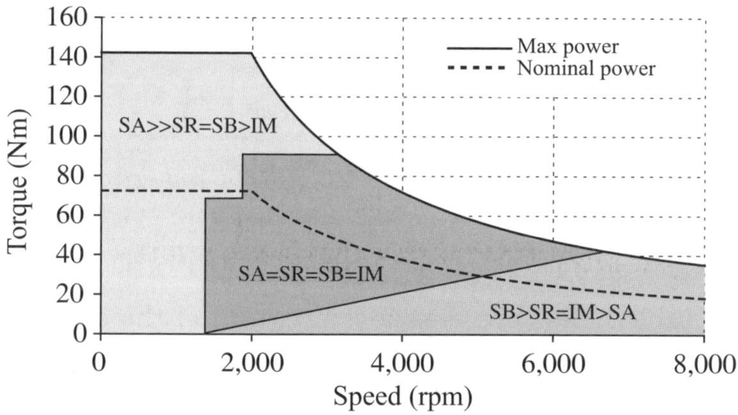

SA: Synchronous permanent-magnet motor
SB: Synchronous wound-rotor motor
IM: Asynchronous motor
SR: Switch-reluctance motor

Figure 4.2. *Efficiencies of different motors*

Electronic structure

The most common architecture corresponds to a *direct connection with the battery* as presented by the diagram in Figure 4.3.

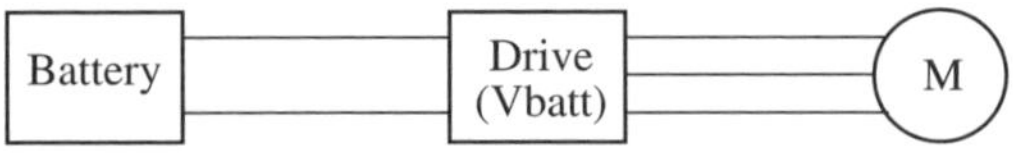

Figure 4.3. *Architecture as voltages of converter and motor are define by battery voltage, all the criterias (like performances and cost) are not always reach*

Another solution consists in add a *buffer DC/DC converter* whose main functions are voltage adaptation of the inverter + motor (choice of the optimized work voltage of the semiconductors, passive electronic parts and engine), and battery decoupling (active filtering, minimization of the requirement in capacitive elements).

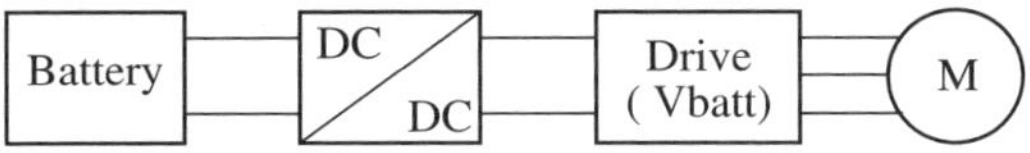

Figure 4.4. *Electric drive with DC/DC converter*

Power electronics

The table below presents a comparison of the currents in power control switches for various motor technologies based on a 30 kW motor fuelled by 200Vdc.

Flux ratio = R	R < 2	R = 4	R = 8
Asynchronous	250A	350A	500A
Synchronous	250A	300A	350A
Variable Reluctance	150A	200A	250A

The currents absorbed on the switch reluctance motor, for the same application, are weakest.

That leads to a less cost of the power inverter parts.

Conclusion

For a purely electric vehicle application, the permanent magnets solution is preferred for many reasons: output in the electric vehicle use area, compactness, mass and noise level.

On the other hand, for a general-purpose use (alternator starter, pure hybrid and electric) and since the cost criterion becomes the determining parameter, the choice of the switch reluctance motor seems relevant compared to the synchronous and asynchronous solutions because of the lower manufacturing cost of the motor and the lowest cost of power electronics, taking into account the weaker control currents for this type of motor.

The cost reductions are estimated at 10% for the motor and 25% for the power control electronics.

However, the use of this very promising technology requires two main challenges:

– control of noises and vibrations;

– reduction of the torque ripple.

Power 30 kW
Voltage 200 CdC (battery
Torque 150 Nm

	DC Motor	AS Motor	Motor syn. bob.	Motor syn. magnet	Switch reluctance
Max efficiency	Good	Fairly average	Good	Fairly average	Medium
Average efficiency	Average	Good	Good	Fairly average	Good
Max speed	Fairly average	Good	Good	Good	Good
PE cost	Fairly average	Fairly average	Average	Average	Good
Motor cost	Fairly average		Average	Average	Fairly average
Torque/speed capability	Average	Average	Fairly average	Fairly average	Good

Table 4.2. *Comparison of electric motor technologies*

4.2. Electronic converters

4.2.1. *Characteristics of electric vehicles*

The use of an electric traction motor in autonomous road vehicles implies the presence on board of an electric energy source, which can be:

– a battery with an appropriate capacity for the desired range of the vehicle in the case of an all-electric vehicle;

– a battery of lower capacity, associated with a range extender (association of a thermal engine and an alternator) or with a fuel cell in the case of a hybrid electric vehicle.

This very brief presentation enables us to notice that in the majority of electric vehicles we will be confronted with the compatibility of DC-current energy sources and alternative-current sources and inside the same category with compatibilities between the voltages of the sources and receivers. This compatibility issue thus implies the presence on-board electric road vehicles of electronic converters whose role will make it possible to remove the operating incompatibilities.

For this reason, we will be able to find on-board vehicles:

– converters of alternating current to continuous current, which we call rectifiers or "chargers";

– converters of DC current to DC current of differed voltage which we call "choppers";

– converters of DC current to alternative current which we call "inverters".

The goal of this chapter is not to give an exhaustive description of each of these part but rather to explain why they are used on-board a motor vehicle.

For readers who wish to have further details on the principles and technologies employed to produce these converters, the bibliography included in the appendix at the end of this chapter should help them to obtain further information.

4.2.2. *Components of electronic converters*

The electronic static inverters which equip electric or hybrid vehicles to modify the electrical energy presentation mode, use associations of electronic power semiconductors, diodes, thyristors, IGBT, which work in commutation i.e. they operate in on or off mode to limit the losses.

In this operating mode, the semiconductor can:

– sometimes be “conducting” i.e. equivalent to a closed mechanical switch which lets the current pass with a direct voltage drop between weak terminals;

– sometimes be “blocked” i.e. equivalent to an open mechanical switch supporting a voltage, between terminals, raised with a virtually-zero leakage current;

– pass very quickly from the “conducting” state to “blocked” state; these are in fact the different changes of state seen at the component or on the circuit, which includes the component that we call “commutation”.

If we take a closer look at the operation of power semiconductors we observe that it is the role they play during commutations which constitutes the main feature of these components. We can distinguish as follows:

– non-controllable components: the diodes which function in the first quadrant of a plane (I, U) having as axes the current as the X-coordinate and the voltage as the ordinate;

– components controllable only with the setting in a thyristor conducting state. These semiconductors can be put in a state of conduction by a control impulse sent onto their "trigger" during an arbitrary moment but they cannot return to the blocked state following a blocking sent onto the trigger. To return to the blocked state, the current crossing them must be cancelled under the very effect of the source or the load, thus causing a "natural" current commutation;

– components controllable when placed in conducting state and in blocked state.

For this type of current the change from blocked to conducting state and vice versa is caused by a current impulse sent onto their control electrode:

– trigger for the GTD and other IGCT;

– gate for IGBT, and field-effect transistors of the MOS type;

– base for the bipolar transistors.

In electric vehicles, the most used semiconductors are diodes, IGBT and MOS. Thyristors are used only in rectifying assemblies.

4.3.3. *Generators – receivers – sources*

In power electronics, receivers and generators are defined with respect to the commutations. Thus it is indicated if it is a generator or a receiver of voltage or DC current or alternative current.

Thus, a generator or a receiver is known as of voltage if the form of the voltage wave on its terminals is not modified by the current which crosses it. A generator or receiver is known as of current when the shape of current wave which crosses it cannot be modified by the variations of the voltage on its terminals.

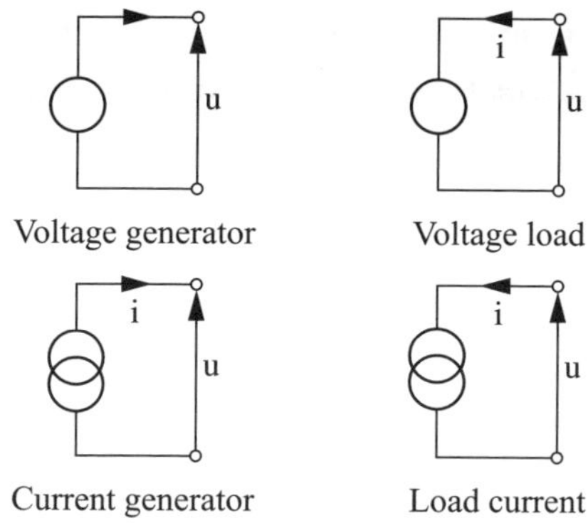

Figure 4.5.

Figure 4.5 gives some examples of the representation of current and voltage "source". Sign conventions for the voltage, U, and current, I, are such that the product U.I has a positive average value for a generator which outputs power and for a receiver which receives it.

Reversibility – sources

A generator or a receiver is also characterized by its reversibility.

Voltage reversibility ($U > 0$ or $U < 0$)

Current reversibility ($I > 0$ or $I < 0$)

Voltage and current reversibility (U and $I > 0$ or U and $I < 0$)

Among the generators and receivers which interest us within the framework of vehicle electric traction, we can observe that a battery and a voltage generator is reversible in terms of current but not voltage. DC motors and synchronous or asynchronous alternative motors are because of their inductance, current receivers; they are reversible in terms of current and voltage.

To also note that as soon as there is a reversibility, the flow direction of the power U.I can be reversed, we then no longer speak of a generator or receiver but a "source".

Correcting, changing the nature of a source

Industrial sources are never perfect and we are led to correct them by the addition of passive components adapted to the type of source considered.

Thus, to correct a voltage source that is too imperfect with regard to the di/dt imposed by the converter, we can connect a capacitor between his terminals.

In the same way when we use a power source that is too sensitive to the dv/dt which are imposed on it, we will be able to improve its behavior by the addition in series of an inductance of a sufficient value.

It should be noted that we can change the nature of the "sources" by addition of passive energy storage elements.

By adding an inductance of sufficient value in series with a "source of voltage" we transform it into "power source"; to transform a current into voltage source it is necessary to connect on his terminals a capacitor with a sufficient value. It should be noted that one can change the nature of the "sources" by means of the addition of passive energy storage elements. By adding an inductance of sufficient value in series with a "voltage source" we transforms it into "current source"; to transform a current source into voltage source it is necessary to derive on his terminals a capacitor whose value of the capacity is sufficient.

Rules of source connection

The connections established by the semiconductors between the input and the output of a converter must respect the basic principles of electrical circuits.

Unicity of the potential in a point

Sum of currents null in a point: the basic principles lead to the rule known as "alternation of the sources". Thus, if at the input of a direct converter there is a source of voltage, at the output there must be a current source, and vice versa.

Characterization of the role of switches

A "switch" is formed by a single semiconductor or a series connection or in parallel of several semiconductors. By grouping several semiconductors of different nature we can obtain switches providing the same function as a single semiconductor but with performance or capacity that we cannot produce with only one component.

4.3.4. Rectifiers

Introduction

Rectifiers are energy converters which transform an alternative-current electric source into a DC-current electric source. We can distinguish:

– diode rectifiers: these converters establish a rigid relation between the output DC voltage and the input alternative voltage;

– thyristor rectifiers which make it possible to vary continuously the relationship between the DC voltage recovered on their output terminals and the alternating voltage applied on their input terminals – while acting on the starting angle of the thyristors;

– rectifiers associating diodes and thyristors;

– groupings of thyristor rectifiers intended to improve performance and decrease disturbances;

– PWM (pulse with modulation) rectifiers which associate IGBT and diodes and which have the advantage of slightly disturbing their alternative input sources.

Use of rectifiers in electric vehicles

In an electric vehicle, rectifiers are used to convert alternative-current electric energy provided either by the public network, or by an on board alternator coupled with a thermal engine; in DC-current electric energy storable in an electrochemical battery and/or super caps.

If energy is provided by the public network, the rectifier system called a "charger" with a single-phase structure. If energy is provided by an alternator, the rectifier is in general three-phase type. Operations of the single-phase rectifier and the three-phase rectifier complete bridge, i.e. using four or six identical switches are described in the following pages.

We can observe straight away that rectifiers, like the majority of converters, are nonlinear operators and their operation results in the injection of harmonics of the fundamental current in the supply source and the consumption of the reactive current. These disadvantages can be combated by means of a positive or active filter, which would exceed the object of this volume to explore here.

4.3.5. *Choppers*

Use of choppers in electric vehicles

In an electric vehicle choppers have two essential applications:

– firstly, they are essential in the supply of propulsion motors when these are DC motors;

– secondly, they are necessary to adapt the main battery voltage to that of the electronic auxiliaries and 12 V network used (sensors, regulators, etc.).

Indeed, we cannot consider the brutal connection of a DC motor on a fixed-voltage energy source – batteries for example – for two reasons:

– no adjustment of the motor torque or speed would be possible;

– the direct electric motor supply would be destructive from both an electric (overcurrent) and a mechanics (overtorque), point of view.

The use of a current chopper located between the energy source and the DC motor makes it possible to solve these two constraints and it brings significant progress with respect to the starting-up processes: insertion of a variable resistor in series between the source and the motor.

The use of a chopper indeed makes it possible to maintain the motor current with the desired value while ensuring the progressive adjustment of the motor voltage without notable loss.

The chopper allows us to regulate the torque and the speed of the motor and thus of the vehicle in traction but also during electric braking.

This braking can be energy recovery, rheostatic or combined – i.e. associated with electric braking and a part of mechanical braking.

Of course, specific provisions must be taken to carry out the controls and protections necessary, for example the use of a filter situated between the battery and the chopper itself to ensure the decoupling of the voltage source consisting of the battery and the traction motor.

Definition

A chopper is a current converter which makes it possible from a DC voltage source (at constant voltage value) to adjust and control voltages and currents, different from the input values and adapted to the needs necessary to supply various receivers (motors, batteries, etc.).

Description

In its basic version, an electronic chopper is composed:

– of a controllable switch (IGBT, Mosfet, GTO, etc.);

– of a freewheel diode;

– of passive components, coils and capacitors, which provide the functions of filtering input and output chopper currents. The mass and volume of these components depend for a large part on the operating frequency of the power switch;

– of controlled device.

Types of choppers

There is way to distinguish between two types of choppers:

– the step-down chopper type e < E.

The average output voltage (e) at the terminals of the load is lower than the average voltage E of the energy source;

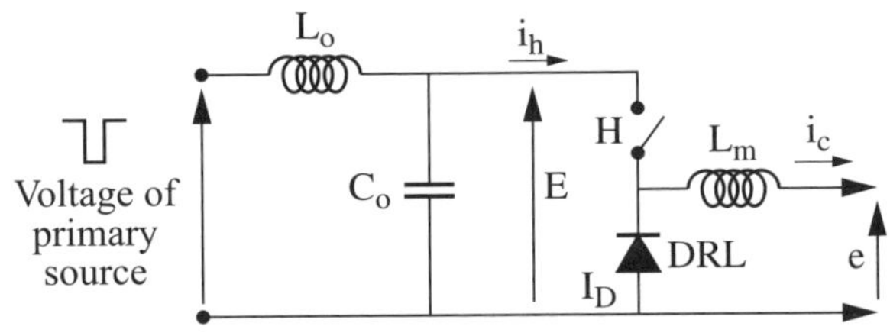

Figure 4.6.

– the step-up chopper type, e > E.

The average output voltage (e) at the terminals of the load is higher than the average voltage E at the terminals of the energy source.

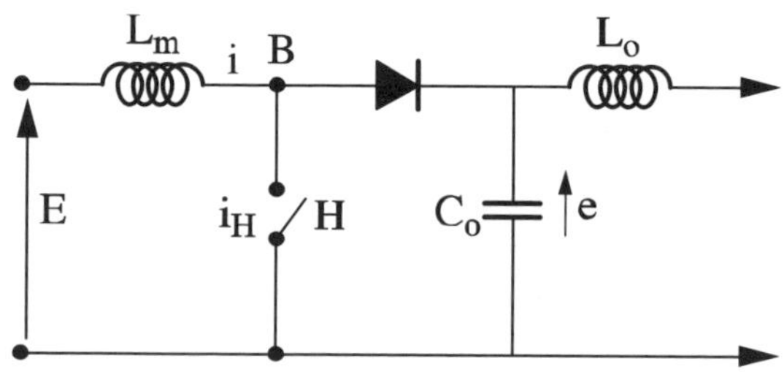

Figure 4.7.

It should be noted that in electric traction equipment, we can pass from one type of operation to another by modifying the topology of the diagram i.e. the position of the switches with respect to the source and the load. We can, for example, work with a step-down chopper in traction and with a step-up voltage chopper in braking to recover the kinetic energy of the vehicle.

The change of topology is done automatically by means of suitable equipment.

Operating principle of a step-down voltage chopper

Let us consider the step-down chopper represented in Figure 4.8 below and the diagram of the currents and voltages:

If we use h to express the chopper conduction ratio:

$$h = \frac{t_1}{t_1 + t_2} = \frac{t_1}{T}$$

E the terminal voltage of the capacitor C_0

i_c the current circulating in the load

e the load-terminal voltage

i_h the current crossing the switch h.

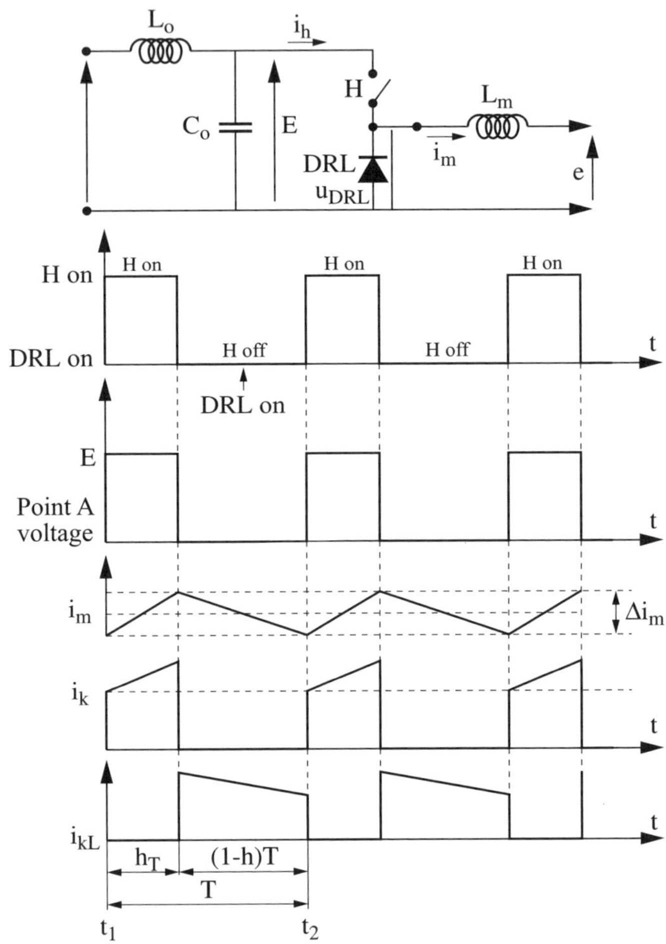

Figure 4.8.

During the conduction time of the switch h we can write,

$$E = e + L_m \frac{\Delta i_c}{hT}$$

During the complementary time h is open and D_{RL} is on:

$$0 = e + L_m \frac{(-\Delta i_c)}{(1-h)T}$$

from which $e = hE$,

if the losses are neglected we can write during the conduction of h:

$$i_h x E = e x i_m$$

from which $i_h = h i_m$,

where the chopper behaves like a DC transformer.

Principle of the step-up chopper

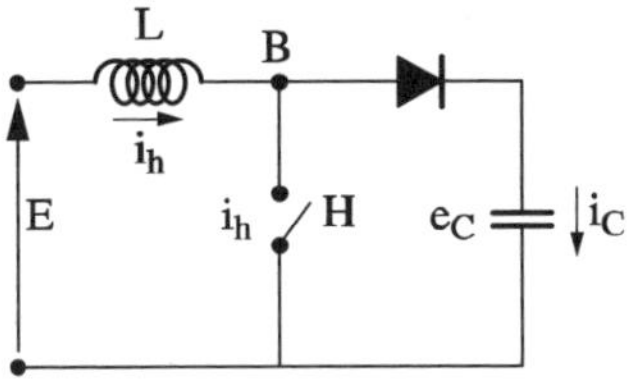

Figure 4.9.

If the DC source has a strong inductance value we can regard it as a current source. Under these conditions, during the conduction of the switch h we can write:

$$E + L\frac{\Delta i_c}{\Delta t} - 0$$

If $\Delta t = hT$

$$\Delta i_c = -\frac{E}{L} hT$$

During the conduction of diode D_{rL} we can write:

$$E + L(\frac{-\Delta i_c}{(1-h)T}) = e$$

or:

$$e = E + .E.hTx\frac{1}{(1-h)T} = E(1+\frac{hT}{1-h}) = (\frac{1-h+h}{1-h})E$$

From which the relation between the different voltages is:

$$e = \frac{E}{1-h}$$

As $0 < h \leq 1 \Rightarrow e > E$

the relation between the currents is deduced from the relation:

$$= E.i_h = e_c \bullet i_l$$

$$i_l = \frac{E}{e_c} i_c = (1-h)i_h$$

$$i_l = (1-h)i_h$$

Interlacing of the choppers

The diagram below illustrates the principle of interlacing of choppers in the case of a step-down chopper assembly.

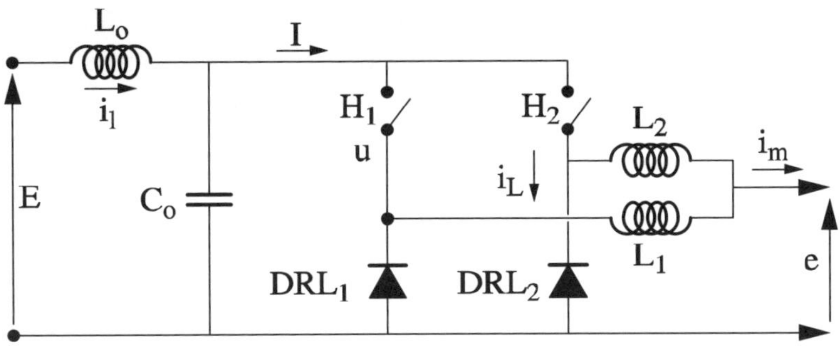

Figure 4.10.

This structure makes it possible to divide the amplitude of the current ripple that the chopper makes circulate on the source line by a factor equal to the cube of the number of interlaced choppers, and to reduce the ripple of the current i_m which crosses the electric traction motor.

Use of choppers in electric drive application

In an electric road transport vehicle, the battery constitutes a source whose value can be regarded as slowly variable, even constant, if we take account the frequency of the chopper semiconductors.

This source has internal impedance which can be significant with respect to the current ripple generated by the chopper operation; for this reason we generally insert a capacitor filter between the battery and the chopper.

This filter has very low impedance at high and intermediate frequency currents.

To find a quasi-continuous voltage at the inductor terminals of the DC motor equal to the average value of the square-pulse voltage delivered at output by the chopper, we use a coil of strong inductance between the chopper and the motor.

This coil can be inductor coil in the case of motor with series excitation – or a appropriate coil if the motor is with separate excitation – and with series image.

The coil plays an essential function in storing the energy which circulates between the battery and the motor. The receiving circuit makes of the traction motor and the smoothing coil can be modeled by a resistance, an inductance and a back electromotive force.

This diagram can be simplified but in a different way according to the kind and the exploitation mode of the motor.

When the motor is with series excitation, and if it functions with full field or with reduced by elimination of inductive turns, the primary current and the induced current have the same value at every moment. The fcem u_m is therefore an increasing function of current i.

It is null when current i is null (without take in count residual current) and proportional to i as long as the motor is not saturated. We can thus represent u_m with a resistance, which is linear as long as the motor is not saturated and which becomes non-linear.

$$u_m = K.N\varphi = R_M i$$

From where the equivalent diagram of the DC electric motor with series excitation for a given speed.

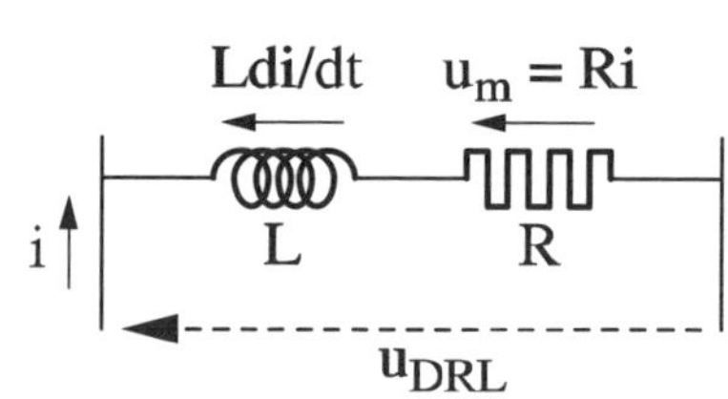

Figure 4.11.

In this diagram:

– L is equal to the sum of all coils: smoothing, induced, inductor;

– R is sum of ohmic resistances, of the smoothing coil, and equivalent resistance of the motor R_M .

When the motor is with separated excitation, the fcem no longer varies according to the current i which crosses through the inductor.

For a given speed and with given we can consider it as constant.

Under these conditions as soon as the speed of the motor rises a little and u_m becomes dominating in front of the ohmic losses in the receiving circuit. We can adapt the following equivalent diagram for the motor:

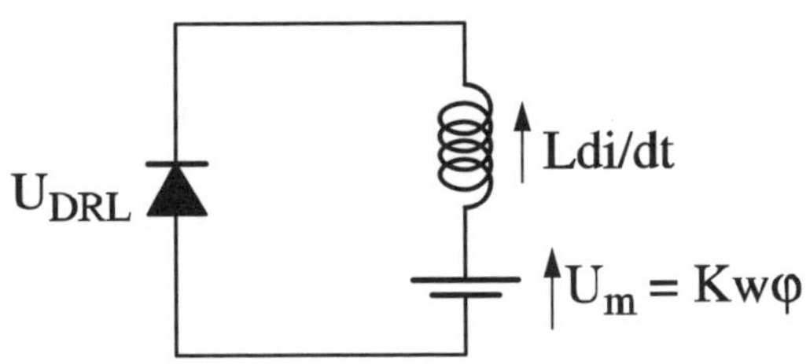

Figure 4.12.

where L is the sum of all the coils of the induced circuit and U_M can be regarded as constant if N and φ are given.

In the case of a motor with series excitation with resistive inductor shunting (either because of a major shunting at high speed or because of a light, permanent shunting). We come back to the case of a separate excitation because the primary current, the flux and the fcem are well smoothed.

In this case we can write $U_M = R_M U_{ma}$, valid only in average value.

Operation in traction with a motor with separate excitation or a motor with series excitation and permanent resistive shunting – calculation of the current ripple

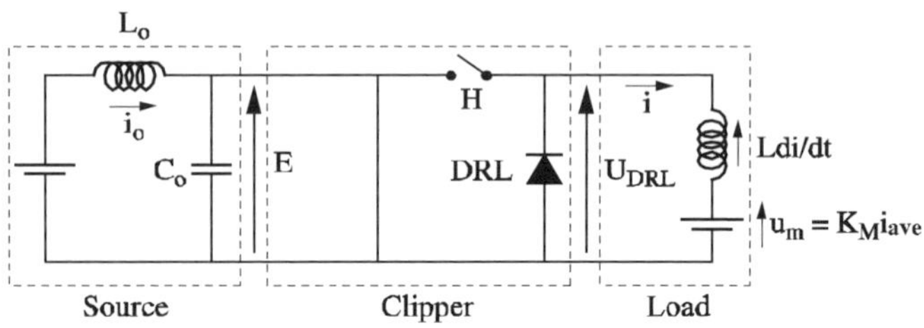

Figure 4.13.

1) In the case where the time constant $\dfrac{L}{R_M}$ is large the current i which circulates does not have time to be cancelled when the switch h is open but the voltage U_{Drl} remains null.

from where $U_{DrLave} = hE$, that is to say: $U_m = hE$

During the conduction of H, we can write:

$$E - U_m = L\frac{di}{dt} = Lx\frac{2\Delta i}{hT}$$

during the freewheel diode time, $U_{DrL} = 0$:

$$U_m = Lx\frac{2\Delta i}{(1-h)T}.$$

from where the value Δi, which is the peak-to-peak half-ripple of the current:

$$\Delta i = \frac{ET}{2L} h(1-h)$$

This ripple is maximum when h = ½ and it then takes the value $\Delta i = \frac{E.T}{8L}$.

We can now plot the following diagrams.

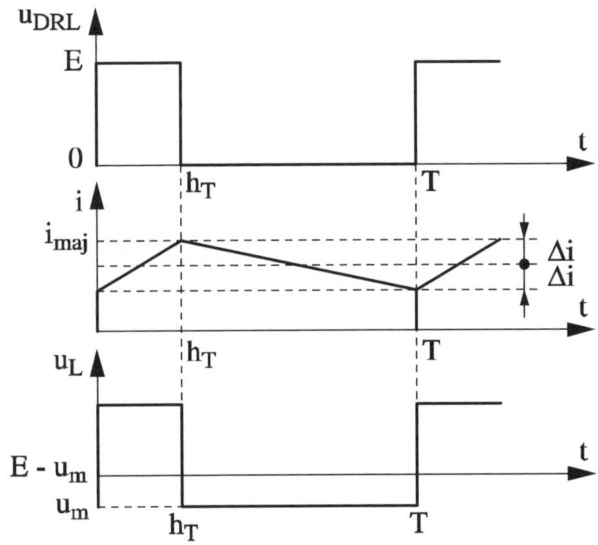

Figure 4.14.

2) If the time constant $\frac{L}{R_M}$ is relatively small, for example in the case where $\frac{2L}{R_M} < (1-h)T$, the current i has time to cancel during the opening time of the switch H. The diode D_{RL} is blocked at the end of the time T_D and supports the reverse voltage *Um*. The diagrams above then become the following:

During the conduction phase of the switch H:

$$E - U_M = L.\frac{2\Delta i}{hT}$$

During the phase when D_{RL} is on:

$$U_M = L.\frac{2\Delta i}{T_D},$$

from where $\Delta i = \dfrac{E}{2L} x \dfrac{hT.T_D}{hT + T_D}$

$$U_M = R_M I_{moy} = R_M x \Delta i x \frac{hT + T_D}{T} = R_M \frac{E}{2L} \cdot \frac{hT \cdot T_D}{T}$$

$$U_M = Ex\frac{hT}{h_T + T_D}$$

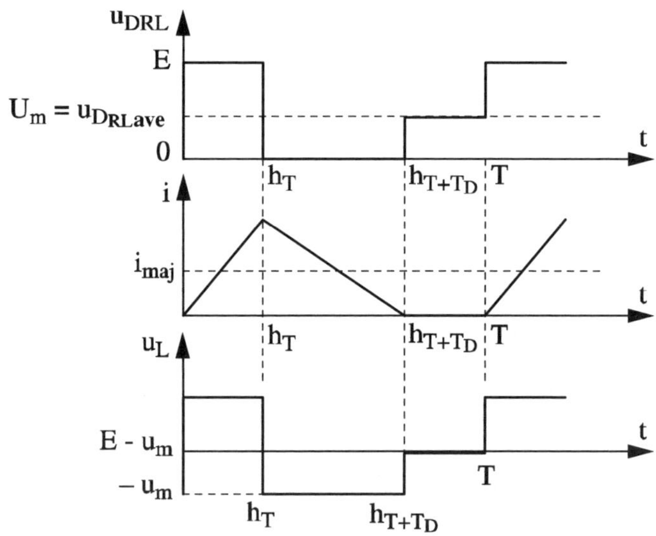

Figure 4.15.

From where, by eliminating T_D:

$$u_M = E\frac{2}{1-\sqrt{1-8\frac{L/R_M}{\mathrm{T}}\cdot\frac{1}{h^2}}}$$

when $h \to 0$ $U_M \to \dfrac{h}{\sqrt{2L/R_M/\pi}}$

Operation in electric traction in the case of a DC series motor without resistive shunting of the inductor

In this case, the equivalent diagram of the electric motor is represented by a coil and resistance in series. The current fluctuations i in the inductor of the motor are exponential in the case where the chopper does not frequently a shutdown phase of the freewheel diode. If we refer to as I and J the maximum and minimum values of the inductor current i and if we decide that:

$$x = \frac{T}{L/R}$$

We can write:

$$\frac{E}{R} - I = (\frac{E}{R} - J)e^{-hx}$$

$$J = I_e^{-(1-h)x}$$

From where the peak-to-peak half-ripple if:

$$\Delta i = \frac{I-J}{2} = \frac{E}{2R} - \frac{(1-e^{-hx})(1-e^{-(1-h)x})}{1-e^{-hx}}$$

This ripple is maximum for h = ½ and we have:

$$\Delta i_{\max} = \frac{E}{2R}\frac{1-e^{-\frac{x}{2}}}{1+e^{-\frac{x}{2}}} = \frac{E}{2R} th\frac{1}{4}\bullet\frac{T}{L/R}$$

Or:

$$\Delta i_{\max} = \frac{ET}{8L}\bullet\frac{th\frac{1}{4}\bullet\frac{\mathrm{T}}{L/R}}{\frac{1}{4}\bullet\frac{\mathrm{T}}{L/R}} = \frac{ET}{8L}\left(\frac{th\vartheta}{\vartheta}\right)$$

In the case where $T << \frac{L}{R}$ i.e. x<<1 $th\vartheta / \vartheta \approx 1$

The exponentials are comparable to lines and we find:

$$\Delta_{i\max} = \frac{ET}{8L}$$

Electric braking in the case of a DC motor

The electric braking of vehicles is of interest for several reasons:

– it makes it possible to recover braking energy in the battery if the latter is able to store this energy, increasing by this fact the electric range of the vehicle;

– it makes it possible to save the mechanical brakes;

– it brings a great flexibility of control because it is possible to combine mechanical brake and electric brake (rheostatic or with recovery).

Moreover, we can recall that the electric braking torque remains limited to a value approximately equal to that of the traction torque. Emergency braking must necessarily combine mechanical braking and electric braking. The respective share of each of these modes can vary according to the speed of the vehicle and the adherence conditions. We can in addition observe that the range increase depends on the mode of exploitation of the vehicle. As an example, recovery on public transport vehicles is optimal when inter stations are long and the speed of vehicles is relatively high, so we can expect to see the rate of recovery on passenger cars move from 10 to 30% according to the mode of exploitation.

The electric braking diagram is particularly simple when we use a DC motor with separated excitation; indeed the inversion of the direction of energy transfer requires only one tiny modification of the topology of the power circuit to reverse the direction of the current which circulates in the battery – and in addition we do not have to be concerned with the value of the residual flux of the motor to start generator operation.

Bridged choppers

We have just presented the operating principles of DC current converters on chopper structures with two one-way current switches. However, when we wish to recover the braking energy, thus reverse the direction of the average current in the battery, we are obliged to modify the topology of the chopper using more or less complex equipment.

This disadvantage can be prevented if a bridged chopper is used. This structure with four bidirectional switches makes it possible to join together each of the two output terminals to each of the two input terminals or to separate them. Indeed, the use of this type of chopper makes it possible to join together the battery which is a reversible power voltage source (E; $i_0 > 0$ or < 0) with a power source – the DC motor – reversible in terms of current and voltage.

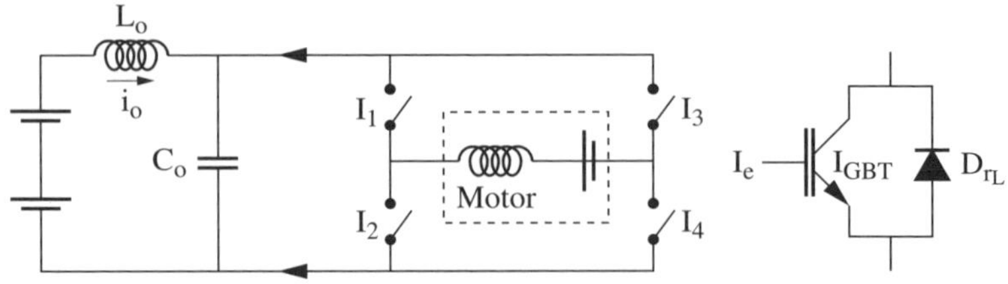

Figure 4.16.

Assembly of a bridged chopper and structure of a bidirectional switch (IGBT + Diode).

By a suitable ordering of the switches I1 and I4 we can find the properties of the step-up and step-down choppers previously described.

4.3.6. *Inverters*

Usage of inverters in electric-powered vehicles

In electric vehicles driven by an AC motor it is not possible to directly connect this type of motor to the continuous voltage source that the battery constitutes. It is necessary to interpose between the energy source and the traction motor(s) a conversion (called an inverter) which transforms electrical DC current into electrical AC current and which makes it possible to carry out the control of the motor torque and the speed regulation of the vehicle both in traction mode and in braking mode.

Continuous-to-alternate conversion can be produced in multiple ways but the characteristics of road vehicles and the rationalization of industrial solutions have resulted in favoring an inverter structure with six bidirectional switches constituted by the association of an IGBT and a diode assembled in anti-parallel and controlled according to a standard PWM (Pulse With Modulation) function. This type of assembly makes it possible to associate a voltage source (like a battery) and a receiver of the power controlled in current (asynchronous motor, synchronous motor wound or with permanent magnet). It should be noted that in a PWM inverter, the output voltage is a succession of voltage pulses (square pulses of voltage) of variable width so that the average value of the fundamental of this wave is sinusoidal over the period of stamping.

This control method has two important advantages:

– it pushes the output voltage harmonics back towards the high frequencies, which facilitates voltage filtering;

– it makes it possible to vary the value of the fundamental output voltage.

The moments of switches' shutdown and opening are generally given in real-time by means of appropriate control electronics.

To explain the operating principle of the PWM voltage inverter, we will initially examine the operation of a half-bridge single-phase current made of two complementary switches I_1 and I'_1 delivering a voltage $V_A - V_O$ equal to $\frac{E}{2}$ or with $-\frac{E}{2}$ and controlled according to a triangle sine modulation.

In this type of control, the moments of control of the switches I_1 and I'_1 are determined by the intersections between a reference wave e_c of frequency f' with a modulation wave M, still called a carrier, whose frequency f' is significantly higher than that, f, of the reference wave. The modulation wave can be of symmetrical triangular form (case of the figure above) – or asymmetrical in the case represented here when f′ >> f, the voltage e_c can be regarded as constant for a period T' of the modulation wave, and the figure above shows that the average value of $(V_A - V_O) = \frac{1}{T'} \bullet \frac{U}{2}\left[T' - (t_2 - t_1)\right]$ is equal to e_c.

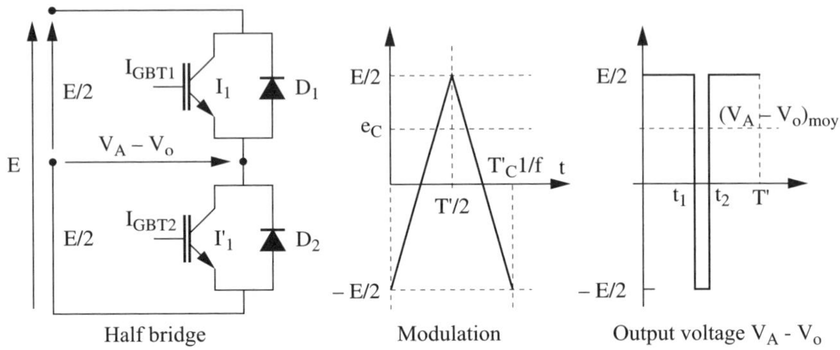

Figure 4.17.

It is thus seen that if e_c has sinusoidal variation, the average value of $V_A - V_O$ varies in the same manner.

In a sinusoidal-type control, two parameters characterize the control of the switches:

– the modulation index $m = \frac{f'}{f}$, which is the ratio of the frequencies of the modulation wave and reference wave;

– the voltage phase-control factor –r- ratio of the control voltage amplitude e_c with the quoted value of the modulation wave.

The modulation is known as synchronous when f' is a whole multiple of f.

Three-phase inverter:

The three-phase inverter with six switches, is obtained by grouping three single-phase current half-bridges described in the preceding section and by shifting their T/D control in time. This inverter is very well adapted to supply AC motors (asynchronous or synchronous controlled in voltage type) with voltage and variable frequency. If we use a sinusoidal reference, the three-phase inverter has the same advantages as the bridge single-phase current inverter, namely:

– possibility of adjustment of the fundamental output voltages;

– rejection towards the high frequencies of the voltage harmonics.

In addition, as the motor is connected in a star or triangle formation without neutral output, there is a removal of the harmonics three and multiple of three present in the output voltage of a half-bridge single-phase current.

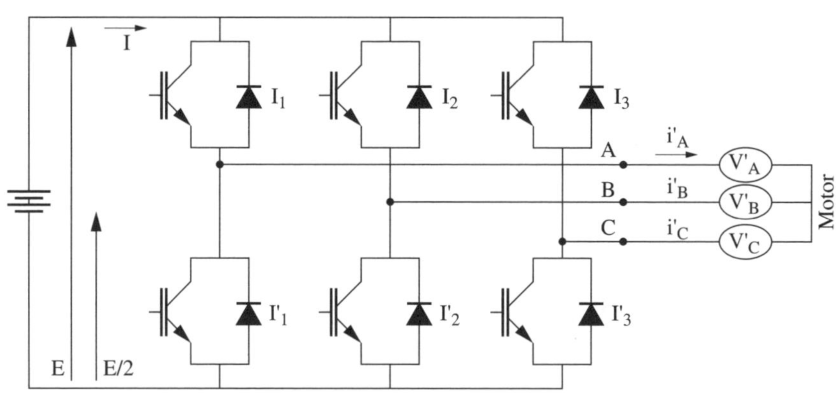

Figure 4.18.

To determine the moments of conduction of the two-to-two complementary switches I_1 and I'_1; I_2 and I'_2; I_3 and I'_3, we proceed as was indicated for the half-bridge single-phase current inverter.

We plot the three reference waves representing the three output voltages desired between the input terminals A B C of the motor and the medium fictive point 0 of the battery. The intersections of these waves with the carrier give the moments of control of the switches. The output voltages are given by the half-bridges:

$$V_A - V_O = \frac{E}{2} \text{ when } I_1 \text{ is closed}$$

$$V_A - V_O = -\frac{E}{2} \text{ when } I'_1 \text{ is closed}$$

$$V_B - V_O = \frac{E}{2} \text{ when } I_2 \text{ is closed}$$

With these simple voltages we can define the composed voltages:

$$U_{AB} = (V_A - V_O) - (V_B - V_O)$$

If the load is balanced and coupled as a star, we can determine the voltages applied at the terminals of the three phases of the motor:

$$U_{BC} = (V_B - V_O) - (V_c - V_O)$$
$$U_{CA} = (V_C - V_O) - (V_A - V_O)$$

Note:

The sinusoidal control is historically the first control algorithm used in PWM; the need to improve the performance of the inverter led designers to work out more sophisticated control algorithms. We recommend those readers interested through their knowledge refer to more specialized works.

4.3. Batteries and static storage systems

This part explains batteries for specific applications like hybrid, electric or plug-in hybrid drive-trains for hybrid and electric vehicles and also for fuel-cell vehicles and the 42V network.

First at all we will focus on technologies with potential now available, then on those which are in the phase of validation, and finally on those which are still regarded as being at research level and which have a possible future for electric traction applications.

4.3.1. *The different electrochemical couples for batteries*

The two different applications: energy and power

The diagram below where the specific energy is a function of the specific power, gives a good representation of the relative performance of the various electrochemical batteries.

It should be recalled that according to the kind of battery, energy or power application, the electrodes will be thinner and the current outputs will be dimensioned more widely in the power case.

Less active material and a greater porosity leads to a smaller specific energy in the case of the power application.

Consequently on this diagram, each couple occupies a larger or smaller area according to its potentiality to offer, at the same time, strong power and energy available.

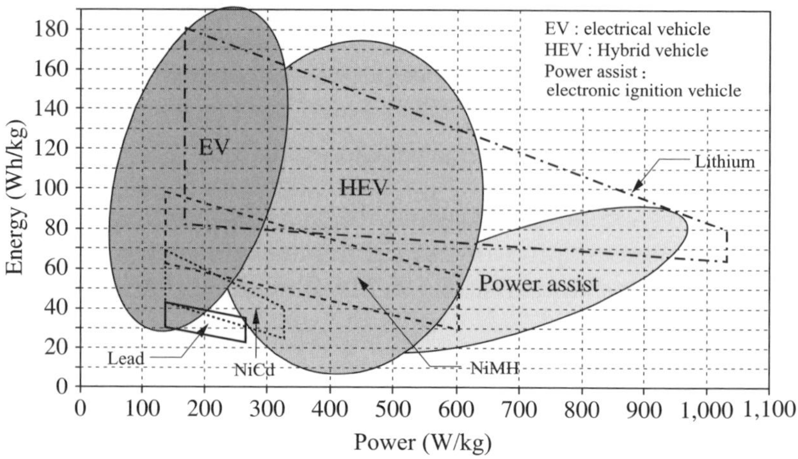

Figure 4.19.

Description of the different battery technologies and state of the art for research in this field

Lead acid technology

Despite all the efforts focused on improving the active material efficiency of the positive electrode and improving its resistance to corrosion, this battery is very limited with respect to this performance for an all-electric or even hybrid application.

Today the innovation comes rather from closed batteries known as AGM, which find an application in stop-and-start or for the 42V supply network. They are characterized by their fiberglass separator in which the electrolyte is absorbed. They have the capacity to cycle until a depth of discharge of 80% and have an operating life 2 to 3 times' higher under these conditions of use. With equal volume compared to a conventional lead-acid battery, the power gain in cold conditions is 30%.

With regard to the electric vehicle application, the performance of the battery is sensitive to the temperature. In addition, its reliability has not been validated. One of the reasons may be a lack of sufficient will from suppliers to provide an effective electronic management system. However, this battery has the advantage of being inexpensive, having an existing recycling process and logistics, and being completely safe. Many stop & start projects use lead-acid batteries; and we can quote the Volkswagen Lupo and the Citroën C2 and C3 Stop & Start which are currently marketed.

Ni-Cd technology

It should very briefly be said that in spite of its robustness and performance that are correct but too small for an all-electric application, the Ni-Cd is unfairly penalized by its bad reputation in environmental terms. Indeed, if we take the example of the electric vehicles which are currently marketed by PSA Peugeot Citroën, the principle of battery rental makes it possible to have a perfect traceability of the battery. In addition, it should be known that the recycling processes exist and that the efficiency of Ni-Cd recycling is more than 99%. Today, this battery seems to have been completely abandoned by all car manufacturers when it comes to research and development work.

Let us nevertheless point out all the strong and weak points of Ni-Cd, excluding environmental considerations. With regard to its advantages, above all it has very good durability. Amongst its disadvantages: an exothermic discharge reaction involving a rise in temperature and ultimately a reduction in power capabilities with the temperature.

Ni-MH technology

Obviously, Ni-MH is the battery of the moment. This technology is now well mastered and largely widespread on the market of portable tools. It is gradually replacing the Nickel/Cadmium battery, by offering increased performance and a better compatibility with the concern of preserving the environment.

The aqueous electrolyte consists of potash, lithia and soda. The positive electrode consists of a nickel hydroxide; the negative electrode is based on the reversible absorption of hydrogen by a metallic hydride. We meet two types of hydride composition currently used: alloys of the type AB2 and alloys of the type AB5.

Japanese and European manufacturers have opted for alloys of the type AB5. These alloys containing a mixture of rare earths need to be allied with cobalt to improve the corrosion behavior and increase cyclability. As an example, Saft uses 5% of cobalt, which unfortunately is relatively expensive.

Ovonic in the USA opted for AB2 alloys. These polyphase materials are rather sensitive to corrosion and give fast degradation over time. On the other hand, they offer the advantage of a stronger slope of the Voltage curve = f (SOC) and thus an easier management of the state of load.

Until now, primarily for production cost and manufacturing reasons, the cells were of a cylindrical design.

What is noteworthy is the introduction onto the market of prismatic cells at Panasonic (PEVE) to equip the Toyota Prius marketed in Europe and in the USA. The announced and the potential for improvement are very encouraging.

Li-ion technology

Lithium generators represent real future prospects for electric powered applications. Lithium is the most reducing element of the classification and has the most negative potential for an electrochemical anode. Its low atomic mass confers to it the highest mass-capacity of all anodic materials: 3.86 Ah/g. Because of its exceptional characteristics, lithium is and will remain the most powerful of anodic materials.

Briefly, the Li-ion system is an electrochemical system founded on the principle of the intercalation and reversible desintercalation of lithium ions in each electrode during the charge and the discharge. The anode is generally made of graphite, and the cathode is a metallic oxide.

The hard reactivity of lithium with water obliges the use of an anhydrous electrolyte, composed of an organic solvent and a lithium salt, much less conducting than an aqueous electrolyte. The development of power elements requires the use of very thin electrodes and separator, which increases the failure risks. Moreover the electrodes and the organic electrolyte can oxidize irreversibly in the event of overload, which imposes an individual control of each element constituting the battery block.

We distinguish three classes of technology according to the type of oxides containing lithium used as materials of the cathode: LixCoO2, LixNiO2 and LixMn204, lithium oxides with cobalt, nickel and manganese, respectively. Manganese oxide has the advantage of being by far the least expensive compared to cobalt and nickel, in a ratio of 20 and 3, respectively. In addition, with regard to safety, the manganese solution seems the best thanks to the stability of the oxide at high temperature. On the downside, manganese oxide is much less capacitive than nickel oxide which itself is less so than cobalt oxide.

To illustrate this, the two Japanese suppliers who develop batteries for electric and hybrid application, Shine Kobe and Japan Storage Battery (JSB) use manganese oxide as a cathodic material. As for Saft, it chose the intermediate way with nickel by combining it (15%) with cobalt to improve its performance and its operating life. The role of aluminum, with a content close to 5%, is to stabilize the system from a thermal point of view.

Li-polymer technology

Li-polymer is as yet only at the development stage. It has a very high potential, but development is still necessary. Under the name Li-polymer, we in fact distinguish two types of technology:

– Batteries with liquid organic electrolytes included in a polymeric matrix or gel electrolyte which is based on a "Bellcore" patent developed by Delphi. The anode is made of vanadium oxide, and the cathode made of graphite. It is what we can call a pseudo Li-polymer. This technology works at room temperature, whereas the batteries containing solid electrolyte must be maintained at a temperature ranging between 60 and 80 °C to obtain a sufficient conductivity of the electrolyte.

– Batteries with solid electrolytes containing ionic conducting polymers which are developed by Argo-tech (Hydro Quebec) and EDF/Bolloré. These batteries containing solid electrolyte must be maintained at a temperature comprised between 60 and 80°C to obtain a sufficient conductivity of the electrolyte.

The Argo-tech and EDF/Bolloré technology calls upon techniques of thin-film lamination and lends itself well to mass production at a low cost. The battery is obtained through the superposition of five materials: an insulator material, a strip iron of lithium for the anode, a solid electrolyte, a metallic oxide for the cathode, vanadium oxide and a current collector.

Super capacitors

Super capacitors are power sources of great interest with regard to operating life (several hundred-thousand cycles), and safety (up to 10 kW/kg). However, the specific energy is lower than 5 Wh/kg, which limits them to only being considered as an auxiliary source of power. Their Achilles' heel is obviously their cost. As a comparison, today at the stage of maturity where the technology finds itself, it is necessary to count a cost eight times' higher than a conventional 12V lead-acid starter battery.

For the moment this technology has not yet convinced car manufacturers primarily of its cost and still remains confined to niches such as military applications, truck starter battery applications and low-emission hybrid buses applications.

4.3.2. *Positioning of Ni-MH and Li-ion batteries for different applications*

High energy batteries

Main results of different technologies:

Technology	Specific energy* (Wh/kg)	Operating life (ECE15 cycles)
Pb	25	< 180
Ni-Cd	40	> 1,000
Ni-MH	60	> 1,000
Li-ion	110	500 < ddv > 900

* Performance on battery system

This table reviews the different technologies, taking into consideration their performance.

Nickel high energy technology

Compared to Ni-Cd, while preserving the robustness of the latter, this technology eliminates the disadvantages related to the environmental considerations relating to cadmium and improves the specific energy by approximately 50% compared to Ni-Cd.

We note a loss of capacity close to 10% at a temperature of -20°C. The cold performances are degraded a little: a loss of capacity can be noted close to 10% at the temperature of -20ºC. From a safety point of view, the results are good in comparison with the Li-ion.

The only constraint is a good ventilation to ensure the dissolution of hydrogen in the air in the event of abusive conditions.

High-energy Li-ion technology

This lithium technology confirms its potentiality in term of performance but show that many technical points still remain to be dealt with before having a completely accomplished technology.

The power objectives are achieved since the values obtained are higher than the objectives, 350 W/kg whereas the objectives fixed the limit at 330 W/kg. With regard to energy, on the other hand, the results obtained are below the objectives, 135 Wh/kg compared with 170 Wh/kg.

We also noted a poor cold chargeability, which prevented us exploring at lower temperatures.

The Li-ion battery system still expensive and even if the performance is improved with regard to both the power and the range, a this price it still remains a penalizing element for the development of the market.

Nickel high-power technology

This system has shown a high potential both in terms of the specific power available and the operating life.

The cell design is cylindrical or prismatic. Compared to the cylindrical format, the prismatic one has the advantage of a gain in volume of 30%, a weight saving of 20%, a power gain of 60% and an operating life improved by 50%.

Lithium-ion high-power technology

In terms of safety, we can say that even if the electrochemistry is the same as for the high-energy cell, the design of the cells is different, a lower amount of energy, and from the operating conditions, cycling around a charge state of 50%, this point is consequently less problematic. All the same, it still has to be improved.

Characteristics	Ni-MH	Li-ion
Specific power (W/kg)	500 – 800	1000 – 1200
Power density (W/l)	1500 – 2,400	2500 – 3000
Specific energy (Wh/kg)	40 – 55	40 – 60
Volume energy (Wh/l)	130 – 180	90 – 135
Cycling operating-life	250 000 cycles	+ 200 000 cycles

Comparison of performance for power Nickel and Lithium batteries

4.3.3. *Recycling processes*

Lead-acid technology has an operational recycling process of course.

Ni-Cd as for them is recycled easily. It is important to recall that the recycling of cadmium is done with efficiency close to 100%. At the end of the operating life, Ni-Cd are collected by networks set up by car manufacturers. Once the batteries of the vehicles, they are sent to a specialized company; in France for example, it is at the S.N.A.M. After disassembly of the battery system, the electrodes are extracted to be recycled. They are treated to extract the cadmium and other materials from them. The cadmium thus recovered is re-used to produce new electrodes, whereas the residue product including nickel-iron is re-used for the production of stainless or special steels.

It should be noted that the industrial batteries delivered in the European countries who adopted the 1991 European directive are recovered almost completely.

With regard to the recycling of the NI-MH, the hard point remains the recovery of hydrides and to a lesser extent the valorization of cobalt. Hydrides are not easy to isolate. In addition the recovered powder, a mixture of various metals does not correspond to the specifications for raw materials bought for battery production. The recovery of hydrides would require heavy investments to set up a complementary process, for which the outlets are not currently assured.

Concerning Li-ion, we can mention that these recycling studies are currently being evaluated.

Conclusion

First of all, a point which it is important to highlight is that in the field of batteries, the evolutions are always slow because of complexity of the chemical systems implemented.

To guarantee the success of a technology in automotive applications, it is essential that they answer the three recurring criteria: cost, safety and performance.

Today the most mature amongst the battery candidates for power applications is incontestably Ni-MH. It correctly meets the needs of the hybrid vehicle and it is already available today.

With regard to the power Li-ion, we can advance that the performance obtained is good and even very good. On the other hand, efforts must be made to improve the operating life and safety.

For all-electric vehicle applications, we find Li-ion and Ni-MH. Li-ion seems to be able to achieve the goals laid down in term of performance at room temperature. However, efforts must be made relating to the improvement of the cold performance and safety.

With regard to Li-Ion batteries, the minimum requirements in terms of power, operating life and safety are assured. Electric vehicles equipped with Li-Ion batteries can be marketed today.

The battery lithium-polymer is very promising from its high performance and its potentially reasonable cost. It currently only exists at the development stage.

However, because of its need to ensure a high temperature, its use for electric traction applications remains problematic.

With regard to super capacities, if their cost decreased appreciably, it could be a possible auxiliary power source.

In addition, the recyclability aspect is not to be neglected, as it is extremely probable that in the future the qualification of a new battery will impose the existence of a recycling process.

4.4. The fuel cell and on-board fuel storage

4.4.1. *History of the fuel cell*

Electric phenomena have forever appeared in nature but it was only in the middle of the 19th century that two important discoveries made it possible to control the use of them:

– the first, in 1831, when Michael Faraday understood that we could generate electricity using the displacement of a magnetic mass close to a conductive wire, thus connecting mechanical work and electrical energy;

– the second, in 1839, when William Grove, a Welsh judge and amateur of science in his spare time, discovered a simple way of directly transforming an electrochemical energy into electrical energy using a device which he called a "gaseous voltaic battery".

Neither were to suspect the consequences of their discoveries.

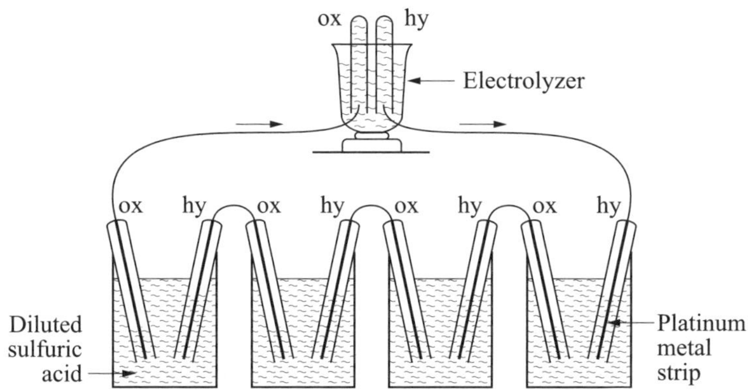

Figure 4.20. *The William Grove fuel cell (four cells)*

William Grove observed the phenomenon which was to lead to the discovery of the fuel cell by undertaking experiments on water electrolysis, proceeded by which water is broken up into hydrogen and oxygen by plunging-in two platinum electrodes connected to a DC current generator. He noted that by replacing the generator with a galvanometer, with the electrodes each remaining locked up in a test-tube containing gases generated by electrolysis, the latter indicated the circulation of an electric current. He had thus recomposed the water broken up by electrolysis. Later, in 1842, he published a paper clarifying the technological problems posed by his invention (properties of the materials, gas-electrode-electrolyte interactions); these problems appeared to him to be so complex to solve that he became discouraged from continuing his work on the subject.

This work was resumed by A.C. Becquerel (in 1855) and Germans L. Mond and C. Langer (in 1889), who baptized the process "fuel battery". As this system is composed of a lot of cells, the name became "fuel cell". Nevertheless, these researchers in turn decided to give up their developments because of the cost of platinum and the difficulties in preventing the poisoning of the electrodes by the carbon monoxide (the hydrogen used at the time came from the decomposition of coal).

It would seem that the first fuel battery designed to generate electricity was produced in 1896 by W.W. Jacques, who used phosphoric acid as an electrolyte.

At the end of the First World War the first fuel cell battery was produced with molten carbonate functioning at 1,000°C, but the fast development of spark-ignition engines around 1930 slowed down the continuation of research on this battery.

Nevertheless, important advances (in particular on the alkaline process system) are to be credited to F.T. Bacon, from 1932.

It was necessary to await the beginning of the 1960s for the real starting-up of the fuel-cell battery owing to NASA's needs on space applications.

They were developed and used in the Gemini, Skylab and Apollo missions with a power of about 1.5 kW; this same technology is still in service in space shuttles (power of 7 kW).

At that time, their excellent reliability had some thinking that it would be possible to use them for road applications, but their cost proved to be too high.

Then the first oil crisis of 1973 caused a resumption of public financing for new energy technologies, in America initially, then in Japan and Europe, restoring hope amongst advocates of the fuel-cell system.

Many demonstrations then followed one another in the various fuel-cell systems, up to powers of 4.5 MW (PAFC type), the most noteworthy players being the American-Japanese ONSI[1] (who sold more than 200 PC-25 200 kW fuel-cells) and the Canadian Ballard (PEMFC type).

This multiple work made it possible to prove the "technological" feasibility of the various fuel-cell systems: only "economic" feasibility remains still to be proven today, which explains the R&D work carried out in all research laboratories, in particular for automotive applications, which represent the most difficult challenge to meet.

4.4.2. *The different fuel-cell technologies*

It will be further on that there exist various types of fuel-cell system but they all keep the following joint characteristics:

– they primarily consist of two electrodes separated by an electrolyte;

– on an electrode (the anode) we start an electrochemical oxidation reaction of a hydrogenated compound (the fuel). This reaction produces electrons and brings into play an ion;

– on the other electrode (the cathode) we start an electrochemical reduction reaction in the presence of oxygen

1 ONSI is now UTC Fuel Cells company (PureCellTM200).

and electrons that have come from the anode via the external load, and which brings into play the same ion as at the anode. Combustive oxygen is generally the oxygen in the air, nitrogen not constituting a poison for the fuel-cell system;

– the electrochemical reactions require a catalyst on each electrode; in addition they produce heat and water;

– between these two electrodes, an electrolyte allows the transfer of the ion concerned while being the electrical insulator with respect to the electrons generated at the anode. According to its nature this electrolyte is liquid or solid.

Several electrochemical reactions are possible: they differ by the type of ion (anion or cation) concerned and by the necessary temperature, which is according to the reaction (in the range of 50 to 1,000°C); it is thus necessary to choose, for each type, the electrolyte which is appropriate.

In fact, six different fuel-cell systems are listed today: they appear in Table 4.3.

Note: the meaning of the acronyms is as follows: AFC (Alkaline Fuel Cell), PEMFC (Proton Exchange Membrane Fuel Cell), DMFC (Direct Methanol Fuel Cell), PAFC (Phosphoric Acid Fuel Cell), MCFC (Molten Carbonate Fuel Cell), SOFC (Solid Oxide Fuel Cell).

Note: in similar tables provided by various sources, two other columns are often added, relating to the efficiency and the range of electric power output:

– the efficiency given does not have any significance as users choose the value which interests them for their application, from the null current value (more than 80% efficiency) up to the full load value for which the efficiency becomes very low. In practice, values ranging between 50 and 55% are generally selected, but they can be different, in particular when the fuel cells are integrated into electric-thermal co-generation systems controlled on request;

– for each type of fuel cell, the range of power is currently widening progressively with the multiplication of the potential applications. Nevertheless, it appears that for small fuel cells (several W to several dozen W), we select the PEMFC and DMFC type, while the PAFC and MCFC remain confined with to powers in the range 100 kW – 5 MW. SOFC fuel cells are seeing their scope of application widening on the other hand, in the range 1 kW – 1 MW. The applications of AFC fuel cells currently remain limited to space and some automotive applications (on-board battery chargers).

Type of fuel cell	**Electrolyte**	**Nature of the electrolyte**	**Ionic type concern ed**	**Temp. (°C)**
Alkaline (AFC)	Potassium (KOH)	Liquid	OH^-	70 – 90
Solid polymer (PEMFC)	Ionic polymer conductor (proton)	Solid	H^+	70 – 90
Direct methanol (DMFC)	Ionic polymer conductor (proton)	Solid	H^+	70 – 110
Phosphoric acid (PAFC)	Phosphoric acid (H_3PO_4)	Liquid	H^+	150 – 210
Molten carbonates (MCFC)	Molten salts (Li_2CO_3/K_2CO_3)	Liquid	CO_3^{--}	550 – 650
Solid oxides (SOFC)	Yttrium ceramic (ZrO_2/Y_2O_3)	Solid	O^{--}	700 – 1,000

Table 4.3. *The different types of fuel cell*

4.4.3. *The PEM fuel cell*

This type of fuel cell is most widespread and it is the subject of the greatest development and the greatest number of applications, in particular in the field of cars.

This observation is the result of several parameters, the most significant of which are:

– the fact of having a solid electrolyte, which ensures better prospects for mass industrialization at low cost and the least risks of corrosion by the electrolyte;

– a level of operating temperature enabling it to start at ambient temperature;

– the capacity to reach a significant energy density and to support high load variations.

Operating principle of cell components

The anode

Pure hydrogen is introduced into the anodic chamber under a pressure within the range 1.2 to 3 absolute bars; this hydrogen is uniformly distributed on the reactive surface by means either of grooved structures engraved in the anode support, or of metallic foam. A carbon felt is pressed against this distributor and plays the function of an interface with the electrolyte membrane. The felt section located at the contact with the membrane is treated as the area of the electrochemical reaction; it is called *catalytic area* in the figure below, and sometimes "*active area*". It is in this area, of a thickness of about ten microns, that *the triple points* are located, i.e. the area of the electrochemical reaction where the catalyst and the ionic and electronic conductors coexist. This catalytic area must meet several objectives:

– to have the largest possible developed surface in order to have the greatest possible active surface, typically several hundred m^2/g;

– to minimize the quantities of platinum concerned by distributing the particles as uniformly as possible (several dozen nanometers) while associating them with electronic conductor particles (carbon particles) and with ionic conductor particles to ensure the conditions of electrochemical reaction and ionic continuity with the electrolyte. Between the 1980s and 1990s, research work made it possible, with equal performance, to reduce the quantities of platinum necessary from a few g/cm^2 to 0.05 mg/cm^2.

Note: this minimization of the quantity of platinum had a disadvantage: that of making the platinum sites more sensitive to the CO molecules which can possibly be present with hydrogen (in particular when it is resulting from a reformer) and which are a poison inhibiting the electrochemical reaction. This situation thus obliges one of the following solutions to be implemented so that the activity of the platinum sites remains correct:

– either to ensure that the quantity of CO present in the hydrogen is lower than 10 ppm;

– or to add a CO oxidation catalyst to the platinum, ruthenium being most effective, to support 100 ppm of CO instead of 10; the associated problem is a raised cost for the electrode, added to the toxicity of this noble metal;

– or to go up in temperature; indeed, it has been proven that the fact of moving from 80°C to 130°C decreased the activity of CO on platinum so much that the anode can support several hundred ppm of CO without any consequences: this way is probably the solution of the future. It will oblige the use of an electrolyte having the best temperature resistance (hybrid structure, for example) but will allow, on the other hand, an easier evacuation of the heat released by the reaction.

The reaction which occurs at the anode is the following one:

$$H_2 \rightarrow 2H^+ + 2\,e^-$$

The electrons thus produced leave the anode to join the external electric charge, the electrolyte not being electronically conductive. The H+ protons, meanwhile, go into the electrolyte and cross it.

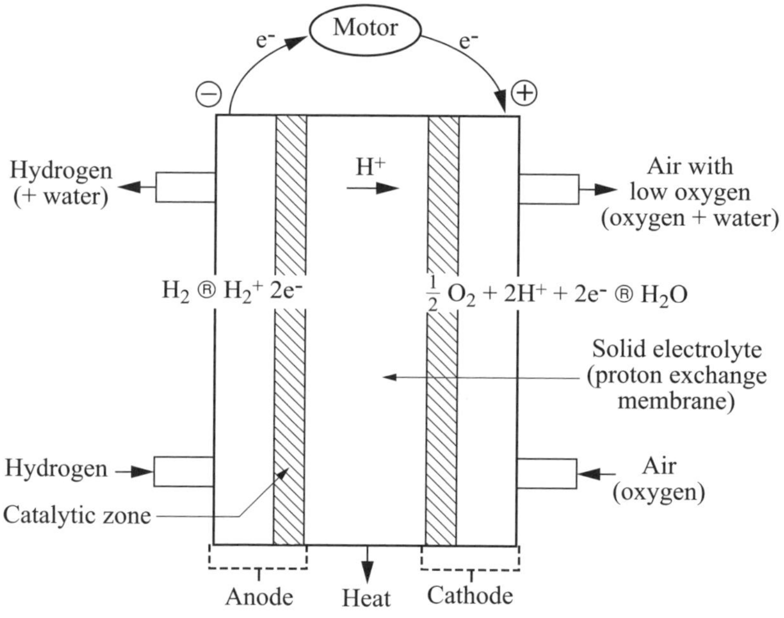

Figure 4.21. *Operating principle of a PEM fuel cell*

The electrolyte

This protonic conducting membrane, the electrolyte, was the key element of the development from the 1980s onwards of this fuel-cell system. Indeed, many properties are required of this membrane:

– to have the highest possible protonic conductivity, in order to limit the losses associated with the decrease in voltage due to the internal resistance;

– to have a good chemical behavior in this acid medium, on a scale of a few thousand hours for automotive applications (several tens of thousands of hours for the stationary applications);

– to offer good thermal properties (in the range 80-100°C) and mechanical properties combining resistance and flexibility, allowing:

- the possible variations in pressure between the anode and cathode to be supported;

- the differential expansions to be supported during power climb and transients, as well as the assembly operations;

- a perfect water tightness between the hydrogen (anode) and the oxygen (cathode) to be ensured.

Membranes meeting all these constraints have required many years of development. Today they generally have the following specifications:

– they have a thickness of about 150 microns when we ask the same material to ensure the performance of ionic conductivity and mechanical resistance, or about 50 to 100 microns when these two functions are separated and we create hybrid structures by incorporating a resistant fabric soaked with an ionic conducting solution;

– they are made of perfluorinated macromolecules containing acid functions SO_3-H^+ whose dissociation, in water, allows the mobility of the protons.

This membrane constituted, even just a short time ago, the final obstacle for an industrial development of the fuel cell because of its cost.

Indeed, it costs, for the most widespread (Nafion® membrane by Dupont de Nemours), approximately €460/m^2, and we need 1 m^2 for 3 – 4 kWe! Recently, this company, and others, have found themselves on the right track to achieve a factor of 20 to 30 on these costs.

The cathode

The cathode, with its area of diffusion and its active area, must meet the same criteria as those of the anode. We generally introduce ambient air with a pressure equivalent to that of hydrogen; only the oxygen part will be useful. This oxygen, accompanied primarily by nitrogen and a little CO_2 will not obey exactly the same laws of diffusion as pure hydrogen, which will lead to a hydraulic electrode structure that is slightly different from that of the anode. Moreover and because of the asymmetry of the flows of the various types concerned, three other parameters must be taken into account:

– the entering air must generally be humidified, because even if the electrochemical reaction leads to the production of water, the input area of the air in the cell is not sufficiently wet to allow the electrolyte to remain a good ionic conductor in this area (see preceding section);

– the oxygen reduction reaction claims more catalyst than with the anode, in order to limit the induced overpressure. In practice, we put around 0.1 to 0.2 mg/cm^2 of platinum;

– the cathode production of water makes it necessary to make the area of diffusion more hydrophobic in order to progressively eliminate water during its production and to thus avoid flooding the active sites. We fulfill this function by a partial impregnation of PTFE.

The reaction which occurs at the cathode is the following one:

$$\frac{1}{2} O_2 + 2H^+ + 2\,e^- \rightarrow H_2O + \text{heat}$$

The returned electrons from the external electric load combine with the protons having crossed the electrolyte and the oxygen present, to produce water; this exothermic reduction reaction occurs in the presence of a catalyst (platinum).

What remains is to eliminate the water and heat. This water is mostly in the form of vapor and also partly in the form of droplets which are pulled by the gas current.

As for the produced heat, it is eliminated in various ways:

– one part is in the form of latent heat in the water vapor and thus eliminated with this vapor;

– another part is eliminated in the form of significant heat, by a rise in the temperature of the liquid part of the water produced and the humidification water introduced in a greater or lesser quantity according to the setting of the operating parameters;

– a third part can be eliminated by thermal conduction towards the bipolar plate which is cooled by an independent circuit (liquid/fluid or gas according to technologies). All the prototypes presented at the beginning of the 1990s presented this function of cooled bipolar plate but solutions have appeared which make it possible to do without it; indeed, this additional cooling system is a source of difficulties (requirement of a specific liquid loop with its peripherals, design of a more complex bipolar plate, requirement of an antifreeze cooling liquid, additional function for the control).

The bipolar plate

The bipolar plate is the physical plate of separation between two consecutive cells. It must provide the following functions:

– *air tightness* to avoid the mixture, and thus the risk of interaction, between the hydrogen of the anode of a cell and the oxygen of the cathode of the neighboring cell;

– *electric conduction* to ensure electric continuity between two consecutive cells;

– *chemical inertia* with respect to the surrounding medium (very acid);

– uniform water tightness; *distribution* of gases across their entire surface (generally via a structure of channels obtained by molding, for example).

The operating features of a cell

For PEMFC they are summarized in Figure 4.22 below.

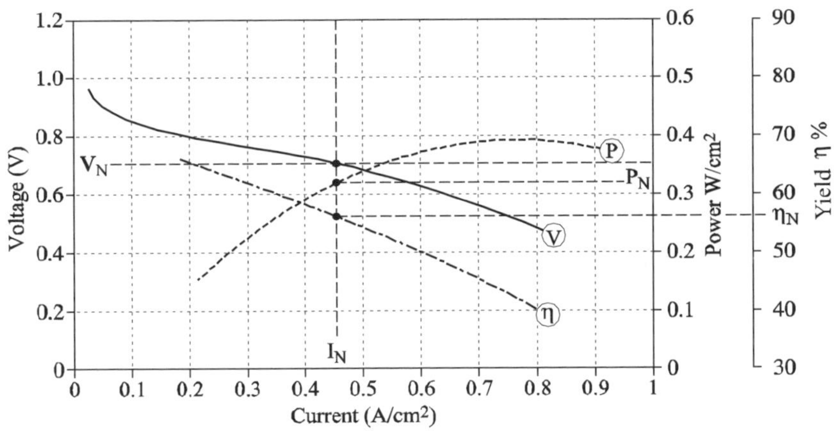

Figure 4.22. *Operating characteristics of a PEM fuel cell*

This type of fuel cell has an operating temperature between 70 and 80°C, imposed by the humidification constraints of the electrolyte currently used (Nafion® type).

We can specify that this current limitation is rather low and that fuel-cell designers would appreciate having an electrolyte which would enable them to operate towards 130°C, which would have several advantages, including:

– greater facility to eliminate the heat produced by a wider temperature variation with the room temperature, thus limiting the size of the thermal radiator;

– smaller influence of CO, a poison to the platinum catalyst, which would thus make it possible to tolerate rates 100 to 1,000 times' greater than the rate currently tolerated (10 ppm of CO).

The basic characteristic of a fuel cell is what we call the polarization (curve **V** on the preceding figure) or current/voltage characteristic. With null current, the terminal voltage of the fuel cell is the difference between the electrochemical potentials to the anode and cathode, that is, 1.23 volt. As the provided electric current increases, the voltage decreases because of the various internal losses that are primarily resistive, and the variation is about linear between 0.8 and 0.4 volt. When the current density becomes high (about 1 A/cm^2), the voltage falls brutally, because of the large increase in the diffusion losses in the electrodes' porous structure and the flooding of the cathode by an excess water production (the quantity of produced is directly a function of the electronic current).

From this polarization curve, we can deduce two other useful characteristics:

– the *efficiency* curve, a function of the current (curve marked η), with the efficiency calculated by the ratio of the voltage at the operating point to the no-load voltage (1.23 V). We will be able to note that the efficiency of a fuel cell is higher than that of a thermal engine owing to the fact that an electrochemical converter does not respond to the same thermodynamic laws;

– the bithermal cycles of Beau de Rochas and Diesel; they obey the Carnot principle for thermal engines, namely:

$$\eta = 1 - T_2/T_1,$$

where T_1 and T_2 are the absolute temperatures of the motor's operations. This theoretical yield can reach 40% for the best motors.

The efficiency of the fuel cell is expressed differently:

$$\eta = 1 - T\Delta S/\Delta H,$$

where ΔH is the reaction and $T\Delta S$ represents the reversible isothermal heat exchanged with the external medium. This reversible theoretical yield is close to 90% at 80°C, which explains part of the passion for this mode of conversion;

– the *power* (U x I), a function of the current (curve marked **P**).

Nominal operating point:

The nominal operating point generally adopted by the users is indicated in the figure above and is at a voltage close to 0.7V. For this voltage, the efficiency is close to 55% and we can note that the maximum capacity that the fuel cell can provide is not reached; this fact is a great advantage of the fuel cell compared with a thermal engine, thus leaving it a reserve of power (about 20 to 30%), making it possible to design a fuel-cell generator for a power lower than the nominal rated output of a thermal engine, the latter then being at its maximum power.

Influence of operating parameters:

The performance of a fuel cell will be affected by the state of the various operating parameters, like the choice of the operating point, the temperature, the pressure, the gas composition, the gas output pressure and the use of reactants, and by other factors like impurities, operating life, etc.

For example:

– while working at a lower current density we improve the voltage and the efficiency of the fuel cell but we decrease its power rating (in terms of both mass and volume);

– while working at higher temperature, we slightly increased the voltage of the fuel cell but we are quickly limited by the mechanical resistance of current electrolyte materials and the constraint of electrolyte humidification. An increase in the temperature has, in addition, a very beneficial effect on the poisoning rate of the electrode platinum by carbon monoxide (CO);

– while working at a higher pressure, we increase the voltage of the fuel cell but we increase the gas compression work and increase the cost of the mechanical structures. Towards the beginning of the 1990s, the operating pressures reached 3 to 4 bars abs. whereas today the pressures implemented are a few hundred millibars above the ambient pressure;

– by increasing the oxygen-cathode ratio (by selective air enrichment with oxygen by making it pass through a membrane, for example), we increase the voltage of the fuel cell but we are likely to be penalized by energy necessary to this enrichment;

– by increasing the air flow to cathode over its stoechiometric value, we slightly increase the performance because of a better oxygen rate towards the output of the cathodic block, but we increase the electricity consumption of the circulator (or compressor) and the sizing of the humidifier. In practice, a flow-rate equal value to twice the stoechiometric value is usual;

– by making the hydrogen recirculate at the anode we more quickly eliminate the water traces and nitrogen traces having crossed the membrane by diffusion, thus increasing the effectiveness of the anode, but we are penalized by the additional equipment (compressor, mixer, purge, control);

– by purifying the hydrogen and air gases, we decrease the influence on the catalyst of the poisons (like CO), thus increasing the performance and the operating life, but the size and the complexity of the purifiers are increased;

– by improving the quality of the components, like the bipolar electrolyte and plates, we increase the operating life but also the cost.

We thus see that the correct choice of the various parameters and factors results from an optimization which will depend on the objectives of cost, operating life, overall dimension and mass. In particular, the technological choices may be very different according to whether we are developing a fuel cell for a light vehicle (3,000 hours of operating life requested) or for a stationary one (50,000 hours requested).

The comparison of fuel-cell and thermal-engine performance

Figure 4.23 below clearly shows that the performance varies very much with the evolution of the load. It is at low load that the difference between a fuel cell and a thermal engine is strongest: we know, indeed, that a thermal engine is less powerful as we move away from its nominal operating point, whereas the fuel cell has an output that is better the smaller its load (except with a very small load where the consumption of the auxiliaries becomes relatively significant).

On the other hand, when the operating point approaches the nominal load, the difference is small: this observation very clearly explains the advantage of hybrid vehicles where we arrange to let the thermal engine function at this point only.

Apart from the application of the thermal engine on the hybrid vehicle, we see that it is essential to carry out the comparison between thermal engine and fuel cell, not on a single operating point but on a complete standard cycle of use: it is not necessary to be devoted to complex investigations to conclude that the energy assessment of a fuel cell will be higher than the best thermal engine (diesel type). Nevertheless, a true comparison should be done only on one complete fuel cycle, “from the well to the wheel”. However, the preceding conclusion would be right only if the fuels were the same ones, which is not the case, except for thermal vehicles fuelled directly with hydrogen like the series of prototypes built by the manufacturer BMW. If thus calculates the overall assessment by taking into account the on-board diesel fuel reforming, the variation in favor of the fuel cell is less. Nevertheless, the following remarks must be taken into account for a complete comparison:

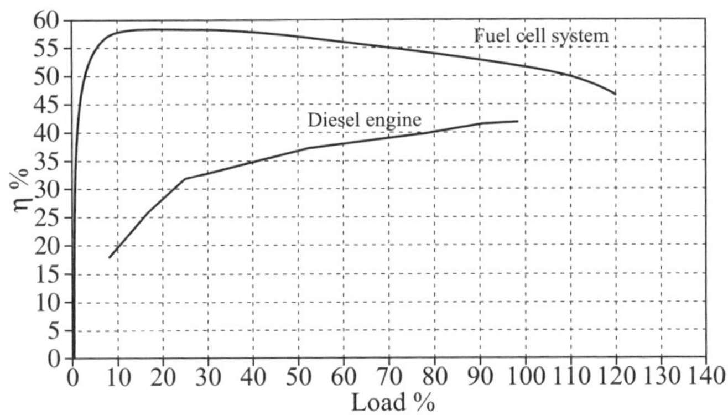

Figure 4.23. *Comparison between the performance of a diesel engine and a PEM fuel cell*

– the only on-board reformers undergoing demonstration are gasoline or natural-gas reformers which thus correspond, not to a powerful diesel engine but to a gasoline engine whose optimum performance is less (approximately 10 points less on the efficiency);

– the comparison and cost analysis will be different according to whether the fuel is reformed on-board or it is reformed in a station to directly transfer hydrogen into on-board tanks;

– the thermal engines remain penalized by their discharges, not always completely treated like gases (CO, NO_x, even SO_2 and CO_2), the micro particles and the noise;

– the fuel cell is a new technology undergoing development and important progress whereas the thermal engine is a mature technology for which the growth potential of the performance is probably more limited.

The advantage of a fuel cell on a motor vehicle is thus less if the on-board primary fuel is from fossil, and only becomes significant when the time come to substitute the current fuel for a non-fossil fuel.

This means that the fuel cell will be essential only gradually in the rate of evolution of the availability and cost of oil and regulatory constraints (progressive imposing of ZEV[2] rules as is the case in the USA, for example).

Note: it should be noted that the reaction time of a fuel cell to a fast modification of the load is extremely short due to the fact of the very great diffusion of gases in the electrodes; we can consider, in the first order, that the behavior in dynamic mode of a fuel cell and a thermal engine are very much comparable.

4.4.4. *Technology and cost of fuel-cell components*

The functions and characteristics of the different components were analyzed previously. Some additional information is given below.

---

2 ZEV: Zero Emission Vehicle.

The electrodes

The anode and cathode are of the same type, namely a carbon felt with a very large, developed surface, less than one millimeter thick, partially impregnated on the membrane side, in the active area, with a water-proofing compound (PTFE[3]) and a mixture of platinum particles deposited on carbon particles, themselves coated with ionic conductor in the area close to the electrolyte membrane; this complex mixture is imposed by the need to ensure electronic and ionic continuities in the electrochemical reaction area (in what is called the active area or catalytic area) and which represents a thickness of around ten microns. They are, nevertheless, not interchangeable because of the different quantities of additives on each electrode: in particular a stronger platinum load and PTFE for the cathode.

The electrolyte membrane

It is more often a question today of an acid membrane of the perfluorosulfonate type: the most widespread is provided by Dupont de Nemours (Nafion®115). It can be described as a skeleton of the Teflon type comprising a sulfonic molecule at the end of the chain (SO_3H). This material needs to be very much humidified to have its best protonic conduction properties and inflates in the presence of water. Its thickness depends on the selected reference (for example, Nafion®115 has a thickness, when dry, of 5 thousandths of an inch, therefore close to 100 microns). This Nafion® membrane have the double characteristic of being most expensive (about €50/m^2) but most resistant (American tests have proved its resistance under operation, without degradation, over more than 50,000 hours). Dupont recently considered the marketing of a membrane whose cost would be ten times' lower, an essential price for automotive application. An alternative, mentioned previously, consists of using fabrics

3 PTFE: polytetrafluoroethylene.

impregnated with conducting polymer fibers (Nafion® type), which makes it possible to decrease the thickness of the membrane and increase its mechanical resistance (for example, the standard GORE® membrane, which can reach a thickness of 20 microns).

The electrode-membrane-electrode assembly (called EME)

This assembly is the unit cell. It is carried out by hot pressing of the three elements described previously, namely the anode, the electrolyte and the cathode. Its characteristics will generally be the following for a motor vehicle, whose available power will be about 50 to 70 kW:

– rectangular frontal surface about 20 x 40 cm, that is, 800 cm^2;

– thickness of about a millimeter, once compressed between its two bipolar plates;

– electric characteristics: 0.7V-400A, that is to say, 280 W.

The cost of the EME is closely linked to the cost of the electrolyte membrane, to which is added today what remains the high cost of producing electrodes, because it is generally done in an artisanal way. They are nevertheless potentially easily produced by automated coating and hot-pressing processes. In the long term, the bottom price will be primarily dependant on the cost of the membrane and platinum (10 to 20 g per fuel cell).

The bipolar plate

All the PEMFC prototypes at the beginning of the 1990s used impregnated, machined graphite plates; they responded perfectly to all the criteria imposed and described previously, but their price (several €100 per plate) excluded this technology from a practical application.

Today this component remains one of the hard points of a fuel cell's cost but a satisfactory solution should soon be available in one (or both) of the following solutions:

– "metallic" way, by stamping a metal plate (some types of stainless steel) allowing the installation of the various openings of the supply and departure of the fluids as well as fluid distribution channels on the pressed plate. This technology is directly derived from that used for the manufacture of heat exchangers plates. Such solutions are currently being tested and should soon be able to reach the cost objectives, which are about €1.5 per plate. An alternative consists of using a metallic, light, flat plate (containing aluminum alloy protected by an anti-corrosion deposit against which the metallic foam is pressed); the difficulty with this method is the installation of an anti-corrosion deposit which is at the same time reliable and a good electronic conductor.

– "polymeric" method, by the mixture of a polymer with a graphite powder in order to give it the desired electric conductivity. This mixture gives rise to a plate of defined geometry, by molding. This method is the most widespread.

In both cases, the thickness of the bipolar plate is close to 1.5 mm and grooves are generally planned for the installation, on the circumference, of gaskets.

Note 1: in technologies implementing a water circulation cooling (with added ethylene-glycol antifreeze), two solutions are implemented:

– two bipolar plates are joined and allow the circulation of the fluid (plate heat-exchanger standard assembly). These bipolar/thermal-exchanger assemblies are then generally set up every 3 to 4 cells; this was the solution adopted by Canadian Ballard and Italian DeNora (which has since become Nuvera) at the beginning of the 1990s;

– the bipolar plate is designed with internal channels.

Note 2: the technology selected will be able to depend on the expected operating life and thus indirectly on the acceptable cost for the stack; in particular, a long operating life (and thus a higher acceptable cost) may lead to a technology very much based on graphite, which is more expensive but more corrosion-resistant.

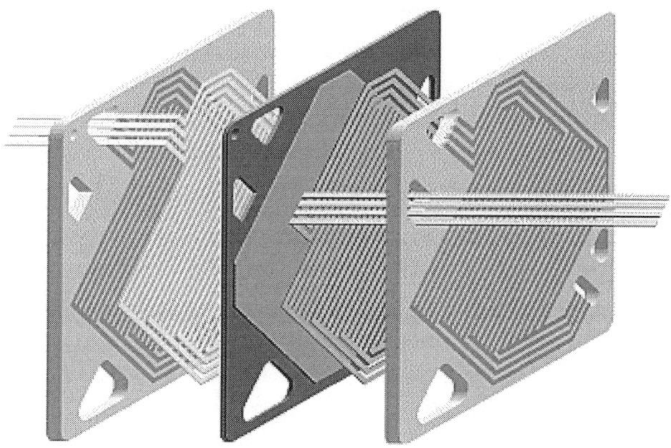

Figure 4.24. *Diagram of a cell surrounded by its bipolar plates (Ballard drawing)*

The stack

The stack is obtained by the electric serialization of a certain number of cells separated by bipolar plates (assembly referred to as *filter-press*). For a desired electric power output, the designer has two parameters:

– the unit surface of the cell which will impose the intensity, with a median value, at the nominal point, of 0.5A/cm^2, that is, for example, 400A for an 800 cm^2 plate;

– the number of cells placed in series which will impose the terminal voltage of the stack, at a rate of 0.7V per cell at the nominal point. In the preceding example, the serialization of 200 cells measuring 800 cm^2 will make it possible to have a source of 56 kW under 140V, with an output current of 400A. It should be noted that such a stack will be able to provide, if necessary, a power of about 65 kW under a weaker voltage, of about 110V.

The dimensions of such a 56 kW stack can be calculated using the preceding data and will be roughly the following:

– length: 200 cells, 2.5 mm thick, to which it is advisable to add the two end-plates ensuring the holding of the cells and the internal pressure resistance, that is, approximately 10 mm each, which as a whole gives a length of 520 mm;

– active section of 800 cm^2 to which it is advisable to add the dead areas necessary for the installation the starting and feeder canals of the fluids, gaskets and the mounting screws, that is, approximately 900 cm^2 of frontal surface;

– volume of about 47 liters, which leads to a specific volume of 1.2 kW/liter, a value obtained by certain of fuel-cell designers like Canadian Ballard in its Mark 900 stack (whose provided installed capacity is 85 kW for a volume of 75 liters);

– mass of about 1 kW/kg.

Under these conditions, such a fuel cell consumes approximately one kilogram of hydrogen for 100km traveled.

Whilst the electric, volume and mass performance has today achieved the anticipated goals, it is not the case for the economic performance: there is habit to give the following figures to explain the situation:

– at the beginning of the 1990s, the cost of developed technology made it possible to reach a value of €5,000/kW, for the stack alone;

– today, the technology available would make it possible to reach a cost of €500/kW, if it was developed industrially, which explains the imminent opening of the stationary market;

– in order to satisfy the constraints imposed by car manufacturers (by comparison with the cost of a thermal engine), the objective to be reached is €50/kW. This value is very ambitious and even appears too much so for those who think that it will be very difficult to dethrone the thermal engine; nevertheless, the technological solutions under development in the laboratories make it possible to realistically envisage it towards the early 2010s. It should be noted that automotive applications are realistic at a higher cost, for niche applications, like an APU[4] to ensure the supply of auxiliaries like air conditioning.

Sustainable technology and life cycle: these concepts are increasingly taken into account in the design of new products. The current components are indeed designed to not implement polluting products (thus, certain solvents in the implementation of chemical components like the membrane and the polymer standard bipolar plate) and others are designed to be easily recycled. Lastly, noble products entering the composition of the stack (such as platinum) will have to be recovered, taking into account their limited availability (and thus their cost). These constraints are taken into account in the design of all the components mentioned here.

4.4.5. *Peripherals of the fuel cell*

Just as a thermal engine is associated with several peripherals to be able to function, a fuel cell is surrounded by the peripherals necessary to satisfy the essential functions. Figure 4.24 shows its layout in a complete system.

---

4 APU: Auxiliary Power Unit.

The humidifier

As was written previously, the electrolyte membrane must imperatively be humidified to be proton-conducting. This humidification can be done in several different ways:

– either the water formed at cathode by the electrochemical reaction is used, bringing back part of it to the entry of the cathodic chamber through the implementation of a capillary structure;

– or the fuel is forced into a humidification chamber where it is almost 100% saturated. In the majority of the prototypes presented in the mid-1990s, this humidifier consisted of several semi-porous plates assembled at the end of the stack, with one side licked by cooling water and the other by the cathodic-volume input air. Subsequently, this humidifier was often placed outside of the stack. In this case, it must be heated by fuel-cell cooling water to a temperature close to the operating temperature;

– or the water is injected at the input of the cathodic volume and the gas is humidified *in situ*.

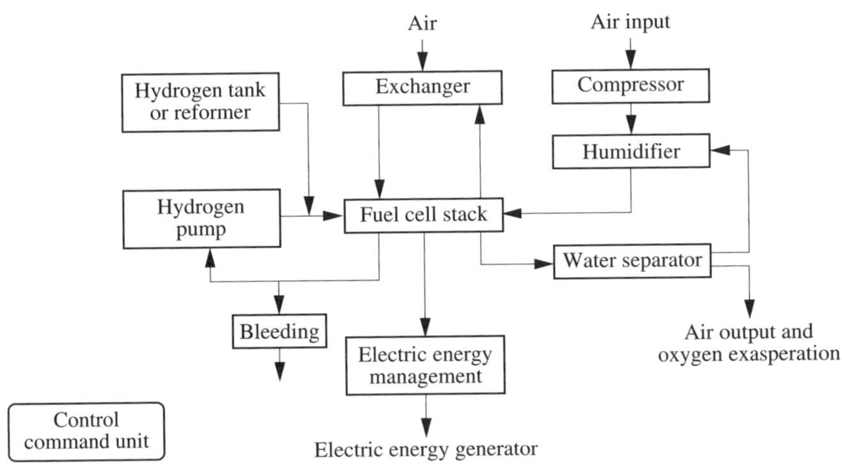

Figure 4.25. *Diagram of a fuel-cell system*

The air compressor

For a long time this peripheral posed many problems, so much so that it is responsible for the operating mode selected today for PEM fuel cells. Indeed, when the problem appeared in the mid-1990s, two methods arose:

– either to develop an air compressor with a compression ratio of about 3 to 4 to bring the air to 3-4 bars, which was the pressure commonly used at the time;

– or to work on the internal thermohydraulics of the anode and cathode to enable them to function correctly at much lower pressure levels.

Practically and curiously, the development of powerful, light and inexpensive compressors did not appear possible and at the same time considerable progress was made on the second method, so that today and according to the pressure levels retained by the designer, it is either air circulators (when the operating pressure is about 1.2 bars ABS) or compressors with a low compression rate (about 2 at the most) which are implemented. They are therefore most generally dry compressors, of the screw type, rather noisy but easy to insulate phonically, which explains the high-pitched whistling which often occurs on prototypes vehicles in demonstration. Such a compressor occupies a volume of about one liter.

The cooling system

By principle of any thermodynamic transformation, the electrochemical reaction is a source of entropy, which is characterized by the production of heat. At equivalent useful output power provided, the fuel cell releases less heat than a thermal engine (efficiency close to 50%, compared to 20 - 40%, according to the type of thermal engine and its operating point), but at an appreciably lower temperature level (around 75°C compared to 90-100°C).

This fact is a handicap because the size of a thermal exchanger (the radiator of a vehicle) is directly proportional, in the most unfavorable case, to the temperature variation between that of the heat to be evacuated and the room temperature. Considering a radiator calculated for a maximum outside air temperature of 50°C, its S_1 surface will have to allow the evacuation of a quantity of heat Q under a variation of 25 degrees, that is to say:

$$S_1 = k\,Q/25$$

Compared with the radiator of a thermal engine of the same provided useful output power and of surface S_2 having to evacuate 20% more heat, that is to say, 1.2 Q, but under a temperature variation of 40 degrees, we see that surface will be:

$$S_2 = k1.2\,Q/40$$

Or $S_1 = 1.33\ S_2$, that is, an exchanger that is a third greater, which is far from being negligible, so much so that the current radiator is already often difficult to integrate into the structure of the vehicle because of the Cx constraints.

This observation explains all the interest of increasing the level of the operating temperature of a PEM fuel cell, which would have the additional benefit – as was mentioned previously, of considerably decreasing the sensitivity of the electrodes with respect to its poisons such as carbon monoxide (CO).

The cooling system is generally a closed loop of water with antifreeze, like a traditional system on a thermal vehicle, but it can also be an open loop by direct water injection into the cathodic volume (Nuvera solution).

The water separator

At the cathode circuit output, we find: air impoverished of oxygen, steam and liquid water. The steam and air are evacuated directly by the muffler but it can happen that we recover part of the water contained in the vapor by condensing it. Water must be retained for two reasons:

– it can be dangerous, particularly in winter, to let water run out on the pavement;

– a certain quantity of water must be deviated towards the air humidifier; the remainder can be stored in a little washer bottle which will be purged in a service station.

The reformer, if there is one, needs water for its operation.

Several fairly simple system types make it possible to fulfill this function, from the traditional "cyclone" to the change-over of the mixture on absorbent porous surfaces.

The electrical energy manager system

As was said previously, the electrical power provided by the stack is delivered under DC average voltage (about 100 to 150 volts) and under high current (several hundred amps). It is necessary to adapt these parameters to the use, i.e. to the electric power drive system, which can be either single or distributed in each road wheel; it must, moreover, be supplied either with DC current, or alternative current, or chopped current according to the type selected, with none the various possible types having, to date, been the subject of a final choice. Lastly, this electric source must supply various peripherals. All these functions for adapting the source to the use are provided by a continuous-continuous or continuous-alternative electric converter according to the technology choices. This converter is controlled by the control unit.

The control unit

All the functions described previously and other secondary functions (such as a purging on the anodic circuit that it is necessary to activate as soon as the rate of impurities reaches certain thresholds) must be managed correctly in the various operating phases, from starting up until shut down, passing by operating phases in established or transitory mode. This function is provided by a control unit which supervises the whole of the parameters available and manages the correlations which must ensure the good behavior of all the peripherals and ensure the safety of the whole in normal or abnormal situations.

4.4.6. *Numerical modeling of the fuel cell*

Reading of the preceding paragraphs clearly that a fuel-cell energy generating system is a complex system in which the interactions between the various parts are multiple in nature: chemical, hydraulic, thermal, mechanical and electric. The development and the optimization of such a system cannot be concluded by empirical methods, no more than from the juxtaposition of optimizations of the various functions taken independently of each other. It is thus justified to proceed with a modeling of each element followed by the dynamic simulation of the unit, validated by many experimental tests.

Several research teams initiated this work at the beginning of the early 1990s, such as the University of Texas and Los Alamos Center in the United States and the "Commissariat à l'Energie Atomique" (Atomic Energy Agency) [SHO 00], in France.

The objective of such a modeling is to make it possible at the same time to offer:

– an understanding of the influence on the system of the modification of one parameter or another, guiding the physical phenomena in question;

– knowledge of the state of the various parameters in any operating point;

– the dynamic response of the system to various requests;

– the elements which make it possible to improve the architecture of the system;

– the elements for an optimum regulation which will help towards the best definition of the control unit operating parameters.

Two types of models are possible:

– *the representation model*, which consists of representing each component using a mathematical function resulting from an experimental knowledge of its operation; the establishment of this mathematical function implies the realization of a great number of experimental tests and limits the validity of this function to the fields actually explored only. The juxtaposition of these functions makes it possible to represent the complete system but with the restrictions previously mentioned;

– *the knowledge model*, which is based on the description of each subsystem by known physical laws, resulting from the fields of implemented (chemical, hydraulic, thermal, mechanical and electric). This type of model is generally less precise than the first type in its fields of validity but requires fewer experimental tests and above all, has a much broader range of use, which is essential for guiding research and development.

The implementation of the knowledge for a fuel-cell system consists initially of breaking it into subsystems, themselves breaking up into other subsystems, etc. The first part of the work consists of drawing up a diagram gathering all the subsystems and specifying the interactions between each subsystem and the material links (flow of energy, flow of material, etc.) and immaterial links (controls, measurements, disturbances, etc.) with each one. Such a diagram will be drawn up starting from a version that is more detailed but similar to that represented in Figure 4.24. To simplify the approach, we will be able to retain some realistic hypotheses, like the constancy of gas pressures at the anode, the constancy of the cooling-water flow-rate, the constancy of the stoichiometry of the air at the cathode, a simple relationship between the hydrogen flow-rate at the anode and the required electric output power.

The most representative parts are modeled in the following way:

– *heat exchangers* (the vapor condenser at the exit of the cathodic circuit, the exchanger between the cooling water and the surrounding air, the cooling exchanger of the condenser, etc.) model themselves very easily using the tools available [CRETh] which, being based on mechanical (type of exchanger) and thermohydraulic data (which lead to pressure-drop and variations in temperature for given heat flows), result in the expression of the physical laws to be taken into account;

– *the circulator or air compressor*, whose representation must be broken up into an electric subsystem (the modeling of which depends on the type of electric motor used), a hydraulic subsystem (which can take various forms according to the type of compression retained) and the thermal subsystem coupled for its cooling;

– *the pressure regulator* at the cathode which can be represented by a simple valve-type system;

– *the fuel-cell system* which can be broken up into four subsystems: an anodic block, a "fuel-cell core" block (which represents the EME assemblies, electrode-membrane-electrode), a cathodic block and a "structure" block. According to the smoothness of the desired modeling, we can break a block up into a certain number of consecutive sections in the direction of the gas flow.

Such a tool was developed by the "Commissariat à l'Energie Atomique" then improved and supplemented in collaboration with manufacturer PSA within the framework of the European HYDRO-GEN program (contract JOE-CT95-0013), the object of which is the study and the creation of a PEM fuel-cell prototype for a light vehicle. In particular, it made it possible to check the feasibility of this type of modeling and the theoretical performance in comparison with the experimental results provided by a prototype test-bench on a scale of a 30 kWe installation. It made it possible, in particular, to provide the component control laws (compressor, regulation valves) and to envisage the behavior of the whole, both in mode in its entire range of possible operations, and in transitory mode, in normal modes (calls and reductions of power) and fault mode. The advantage, ultimately, of such a model compared with a representation model, is that it can be modified and adapted to any technological evolution, which makes it applicable to future prototypes.

4.4.7. *The fuel and its storage*

The PEM fuel cell can only be powered with one fuel: *hydrogen*. In an automotive application, it is thus necessary to have on-board, either hydrogen in a quantity which will depend on the desired range (approximately one kilogram of H_2 for 100 km), or another hydrogenated fuel which it will be necessary to progressively transform into hydrogen as and when needed. Other more exotic techniques exist but are not used yet, such as "Powerball" technology, which consists of making the balls of sodium hydride react with water.

The first two methods are analyzed more detail below.

4.4.7.1. *Hydrogen storage*

The storage of hydrogen on-board can technically envisage many methods; the following parameters will make it possible to compare them and to make choices:

– mass;

– volume;

– response time and inertia;

– filling;

– cost of the storage system;

– level of development;

– possible coupling with an external energy source (thermal or electric);

– safety (intrinsic and accidental).

The best-known modes of storage are listed hereafter and briefly analyzed according to the criteria above.

Gas storage under pressure

The first experiments (the first-generation Ballard and Daimler Chrysler buses) were based on storage at 200-250 bars. The large majority of the current demonstrations are based on a storage with 350 bars (European project of bus CUTE, DaimlerChrysler FCell, Ford FCV, Honda FCX, Nissan X-Trail, etc.) but the most recent models pass to 700 bars (GM Sequel, Toyota Fine-N, etc.).

The materials are most generally made of cylinder-spherical composite structures formed from a sealed bladder made of aluminum alloy reinforced by an impregnated carbon fiber winding made of Epoxy resin.

Nevertheless, metallic structures are also used (Scania FC-Bus project). The major players of these "composite high pressure" technologies are Quantum Technologies Inc (Ca, USA), Dynetek Industries Ltd and EDO (Canada), Ullit S.A. and Composites Aquitaine (France):

– mass: the best current performance is that obtained by Quantum[5], who announced a mass performance of 13% at 350 bars with a safety coefficient of 2.25 (i.e. a tank resistant to 790 bars). The 700-bar tanks developed by the CEA have a lower mass performance (6.5%) because of a higher safety coefficient (2.5, therefore a pressure of 1,750 bars), and aluminum-bladder and carbon-fiber technology based on lower cost and not on the mechanical performance. It is thus seen that such a pressurized tank has a mass always lower than 100 kg for 5 kg of hydrogen on-board;

– volume: the performance of the 700-bar CEA tank is 25 g hydrogen per liter of the tank, which corresponds to a volume of 200 liters for 5 kg of hydrogen on-board. For a 350-bar tank, this volume will be higher. It will, however, be of a lower ratio than the ratio of the pressures, but should reach 50% more, that is to say, 300 liters, which would probably be a decisive handicap for a light vehicle (which explains the 700-bar projects of Daimler Chrysler), but not for a bus which has a significant available volume on its roof;

– response time and inertia: this parameter does not have to be considered with this technology insofar as the hydrogen flow available has the same time-constant as that of the opening/closing valve;

– filling: it is an important parameter for a future network: the higher the pressure, the higher the compression energy and the more the compressor in the station will be expensive;

---

5 *Fuel Cell Technology News*, p. 13, October 2000.

– cost of the storage system: this composite-tank technology is developed in the same way as natural-gas storage because they are very close; they appear compatible, in the long term, with the imposed cost constraints;

– level of development: this technology has been mature at 350 bars since the end of 2000 and is at the stage of full-scale demonstration at 700 bars;

– possible coupling with an external energy source (thermal or electric): this type of storage does not require any coupling with outside, not having any temperature constraints;

– safety (intrinsic and accidental): the intrinsic safety is good insofar as the leakage rate of such a structure is extremely small and is managed without any problem. In accidental mode, many “crash landing tests” were carried out (in particular by the INERIS[6]) and the behavior in extreme situations (fire, falls, impact, shooting with real bullets) seems satisfactory taking into consideration the conventional criteria.

Storage in glass micro-balls

This type of storage is based on the double property of certain types of glass of being tight with respect to hydrogen at room temperature and porous under hot conditions (above 350°C). In addition, the mechanical resistance of glass makes it possible to reach bursting pressures, with hollow spheres, about 1,500 bars with micro-balls around 2 microns thick and around ten microns in diameter. This product is available on the market[7] for other applications (composite material lightening, in particular). In the early 1990s, the CEA tested this material as a hydrogen tank, for automotive applications, which enabled it to characterize the various products available [ALL 99].

---

6 INERIS: Institut National Environnement Industriel et des Risques.

7 From Potters, 3M and Sovitec companies.

– mass: the best results made it possible to reach a rough mass performance of 5.5%, that is to say 55 g of hydrogen per kilogram of balls, with an internal pressure of 1,000 bars. To have the net mass performance, it is necessary to add a thin envelope (it does not need to hold under pressure), heated and thermal insulated, which should make it possible to reach a value ranging between 4.5 and 5% and probably a little higher by treating the balls to make them more resistant to pressure;

– volume: the best performance (in the same conditions as above) is of 20 g of hydrogen per liter of balls, that is to say 250 liters of balls for 5 kg of hydrogen;

– response time and inertia: they depend on the selected system of use. Indeed, two solutions were adopted: either heating to 350°C which has an advantage (flexibility) and two disadvantages (long response time and need for an external energy source), or crushing ("coffee mill" or vibration type) which has the advantage of a very short response time, but the disadvantage of a complex technology and management (crushed glass);

– filling: it practically implies an tank that is interchangeable with filling in the station, taking account of the fact that a compression of a 1,000-bar gas is expensive both in terms of the compressor involved and the energy (it is necessary to spend approximately 30% of the energy contained in the gas);

– cost of the storage system: this cost has not been calculated with precision for a large-series strategy because this solution was quickly abandoned in France, but the cost of the raw material (balls) is quite low;

– level of development: tests in laboratory on a few hundred grams;

– possible coupling with an external energy source (thermal or electric): as written above, this type of storage requires an electric coupling with the outside, whatever the solution implemented, or possibly a mechanical coupling;

– safety (intrinsic and accidental): the intrinsic safety is excellent insofar as hydrogen remains locked up as long as no action is taken on the balls; moreover, at 1,000 bars, tests have shown that there is no risk of a bursting in chain if a ball explodes spontaneously. This fact would ensure good behavior in accidental mode but tests have not been performed to this stage.

Since 1994, the CEA and the manufacturers PSA and Renault have decided to give up this method, thus not allowing us to continue the technical-economic characterizations of this storage method.

Storage in the form of hydride

A hydride is a solid compound formed by direct action (and theoretically reversible) of hydrogen gas on a metal or a metallic compound. The heating of such a solid compound releases hydrogen gas. There exist a very great number of metallic compounds but the most commonly proposed for this type of storage are the following:

– Fe-Ti alloys: they have a relatively small storage capacity (# 1%) but they have a relatively low cost and a low formation enthalpy (<8 kcal/mole);

– $LaNi_5$ alloys which have a slightly better capacity (# 1.4%);

– alloys containing titanium (other than Fe-Ti) on which several laboratories are working, such as those of Toyota and Siemens, and which allow values ranging between 2 and 3% to be reached;

– alloys containing magnesium (as Mg_2Ni) which have a high mass capacity (up to 7.6%) but which have a high desorption temperature (about 300°C) and is very sensitive to impurities;

It is worth noting the lithium-beryllium alloy recently[8] presented by McGill University in Canada with an astonishing performance of almost 9%, the temperature of desorption being only 150°C.

The characteristics are the following:

– mass: being metallic alloys, the densities are generally high. For example, the Ti-beta alloy developed by Toyota and implemented in its RAV-4L fuel-cell prototype, had a mass capacity of 2% (that is, 250 kg of tank for 5 kg of hydrogen on-board but this prototype only had approximately 2 kg on-board);

– volume: the volume capacity is of interest. In the case of Toyota, the value was 37g H_2/liter, that is to say, a volume of 135 liters for 5 kg of hydrogen on-board;

– response time and inertia: no data is available for large-sized tanks, but it is certain that the response time of the tank to an external request is several minutes and that conversely hydrogen continues to be desorbed at the end of the request. That means that a fuel-cell vehicle solution with a storage on hydrides requires a surge buffer tank;

– filling: the hydrogen filling generates a complex environment to ensure the cooling of the structure (the reaction is exothermic) and to make sure of the state of filling, which will depend on the internal structure and the porosity of the alloy. Perhaps, taking into account the time necessary (probably several dozen minutes), it is necessary to imagine a standard replacement of the tank in station;

– cost of the storage system: no data is available but the cost of the basic material is not very high;

– level of development: the development of the hydrides is very advanced for various applications (Ni-MH batteries), but in fact the same materials are not usable, the

8 *Hydrogen & Fuel Cell Letter*, p. 5, July 2000.

specifications being different. As for automotive applications, it is the Japanese in particular (Toyota, Mazda, and Honda) who have implemented this technology on automotive prototypes;

– possible coupling with an external energy source (thermal or electric): as written previously, this type of storage requires several couplings with the outside: a set of sensors coupled with the control unit to control the state of charge and discharge, an electric connection to heat the tank for the discharge and possibly a thermal connection to cool it during charging;

– safety (intrinsic and accidental): the intrinsic safety is excellent as the hydrogen remains imprisoned in the alloy as it remains at room temperature. In accidental mode, the data are not available but put aside from the pyrophoric behavior of some alloys (magnesium in particular), the behavior should be good.

Note: on the initiative of the IAE (International Energy Agency) a database gathering the main properties of 1,200 metallic compounds is freely accessible on the Internet [HYD].

Storage on activated carbon (cryoadsorption)

The adsorption of hydrogen on activated carbon has been known of for a long time and has been the subject of many publications [CAP 94]. Within the framework of its research programs, in the early 1990s the CEA carried out characterizations of carbon that were powerful and available in the trade, like F12 from Chemviron Carbon, TA100 from PICA and Maxsorb from Tokyo Zairyo USA Inc. To be effective, these carbons must be used at the temperature of liquid nitrogen (77 K) and under medium pressure (30 bars):

– mass: Maxsorb carbon made it possible to reach the rough mass performance of 10.7%, which would correspond to a carbon mass of 47 kg for 5 kg of hydrogen, to which it is appropriate to add the weight of the dual-envelope enclosure (liquid nitrogen trimming);

– volume: the same carbon leads to a rough volume performance of 39 g of hydrogen per liter of carbon, that is approximately 130 liters for 5 kg of hydrogen;

– response time and inertia: the response time is a few seconds insofar as the dynamics of desorption is very fast above 77K. Nevertheless, this type of storage should be associated with a buffer tank;

– filling: as soon as the temperature of the carbon is 77K, adsorption is very fast, carbon being a good thermal conductor;

– cost of the storage system: no data is available but the enclosure should have a fairly high cost because of the double enclosure cooled with liquid nitrogen;

– level of development: laboratory demonstration;

– possible coupling with an external energy source (thermal or electric): two thermal connections are necessary, one to maintain the carbon at low temperature and the other to heat it during the discharge;

– safety (intrinsic and accidental): the intrinsic safety is not good insofar as a partial lack of liquid nitrogen causes a very fast increase in the internal pressure.

For the various reasons clarified previously (and primarily for reasons of safety and logistical constraints), this type of storage does not appear to be adapted to a road vehicle and was thus rejected by car manufacturers, in 1994, within the framework of their collaboration with the CEA (VPE/PAC Project).

Storage in carbon nanostructures

This type of storage is without any doubt that which at the present time is exciting the most imaginations. This excitement finds its origin in a communication that Nelly Rodriguez made at the 1996 MRS convention [ROD 96], which led the press to translate his results by affirming that nanofibers could store up to three times their own weight in hydrogen at ambient pressure and temperature! It was enough to imagine a 100kg tank which would contain 75kg hydrogen, to cover 7,500 km without refilling!

The basic idea is to imagine that a hydrogen atom removed of its single electron, which is easy to do, (a proton) was so small that it could easily thread into the interstices of the carbon nanostructures to attach itself to multiple places.

Five years later, optimism dropped significantly owing to the fact that the figures advanced related primarily to experiment errors (taking into account the difficulty in manufacturing nanotubes, the masses concerned in the experiments were about a gram, or sometimes less). Scientists think today that there is an pathway interesting to explore but that we are still very far from the hope of an application, the few credible results (because they were carried out with a mass of at least 100g of carbon) have not made it possible to date to exceed a mass capacity of 1 to 2%.

Note: there exists an Internet site that brings together all the updated on this subject [NAN].

Storage in liquefied form

This type of storage has the advantage of calling upon a cryogenic form that is fairly widely diffused; it has the disadvantage, on the other hand, of being an "energy consumer" since the energy to be spent to bring hydrogen gas from room temperature to its liquid form is about 40% of the energy which it contains. Moreover, the need to compensate for thermal losses necessary to maintain it at 20K causes a permanent evaporation (thus generally a loss) of about 2 to 3% per day, a value that is a function of the quality of the thermo insulation, therefore of the cost of the tank. An alternative (used by Air Liquide on the Fever prototype by Renault and by BMW on its prototypes) consists of pressurizing the tank slightly (5 to 10 bars), which makes it possible to block the leak for 2 to 3 days, and limit the losses but increase the on-board weight.

This type of cryogenic storage was also implemented by DaimlerChrysler on its fuel-cell prototype Necar 4.

– mass: in addition to the hydrogen mass to have on-board, it is advisable to add the mass of the tank. A median value is 70g/H_2 per kg of tank, which leads to a tank of about 70kg for a load of 5kg of hydrogen on-board.

– volume: because of its weak density (70.8 kg/m^3 at 20.3K), the rough volume corresponding to 5 kg of hydrogen is about 70 liters, to which it is advisable to add the volume of the thermally insulated enclosure. A median value is 30g H_2 per liter of tank, which leads to a tank of about 170 liters for 5 kg of hydrogen on-board.

– response time: the response of such a tank to the request is virtually instantaneous;

– filling: the filling of a liquid-nitrogen tank is mastered perfectly; BMW recently showed an automatic filling for vehicles (Munich airport). The filling time is a few minutes;

– cost of the storage system: no data was provided by the only manufacturer using this technique (BMW on its series 7 vehicles, with thermal engine);

– the level of development: demonstration in small series;

– possible coupling with an external energy source (thermal or electric): it is necessary to envisage a small heating resistance within the liquid;

– safety (intrinsic and accidental): intrinsic safety is average insofar as hydrogen leaks either permanently or periodically (via a purging valve); nevertheless, the manufacturer BMW was able to manage this disadvantage in a very sure way. Concerning accidental safety, BMW carried out all the "crash tests" necessary to satisfy German regulations and to obtain the authorization to put its vehicles on the road.

The storage of a fuel to be transformed

With the storage of hydrogen on-board currently posing various problems such as, mainly, the availability of a hydrogen distribution network, many manufacturers consider, at least in an intermediate phase, the solution of having a hydrogenated fuel on-board (hydrocarbon or alcohol) which it is necessary to transform or "to reform" progressively to produce hydrogen as and when needed.

Several processes exist for transforming such a fuel which would be, for an automotive application:

– either a hydrocarbon: gasoline, diesel or natural gas;

– or an alcohol: methanol or ethanol.

Vapor reforming

This process is one of most widespread for the production of hydrogen and one of the simplest. In a first stage, water vapor is mixed, under pressure (a few bars), with fuel and in the presence of a metallic catalyst, generally containing nickel; if a hydrocarbon is concerned, the temperature of the mixture must be about 850°C whereas it should be only 300°C with methanol. The fuel is then oxidized to make CO and hydrogen in an endothermic reaction:

$$C_nH_m + nH_2O \rightarrow nCO + (n+m/2)\ H_2$$

CO is then subjected to a reaction known as a "shift", towards 400-500°C during which it reacts with water to give a complementary quantity of hydrogen and CO_2 in an exothermic reaction:

$$CO + H_2O \rightarrow CO_2 + H_2$$

This set of two reactions is endothermic overall; the heat necessary is brought by the fuel.

The CO_2 can then be separated from the gas mixture by a separation membrane or left in the mixture introduced into the anode of the fuel cell insofar as this gas is not a poison for the fuel cell.

Owing to the fact that this process requires a rather high temperature, the response time can be rather long, in particular at start-up, during which a time of about five minutes is necessary. This fact thus obliges the presence of a buffer tank or a hybrid assembly (the fuel-cell generator coupled to a battery). Another difficulty is avoiding the production of carbon deposits instead of CO. Lastly, it is difficult to avoid CO traces in hydrogen at the exit; the parasitic quantities of CO must be lower than 10 ppm, to avoid poisoning of the fuel cell, which is now realizable in established but also impossible in transitory mode, in particular in the start-up phase, except implementing a filtration (containing palladium on ceramics, for example), which is expensive and fragile.

Partial oxidation

In this process, which is comparable with pyrolysis, we break the fuel molecule under the effect of heat; the fuel is burned in the presence of a limited quantity of oxygen to avoid total combustion. This reaction is exothermic overall.

Reforming known as "autothermal"

The reformers currently developed for automotive applications are generally based on a coupling of these two types previously described so that the quantity of energy required is null, from where its name "autothermal". For example:

$$CH_{1.4} + 0.3\ H_2O + 0.4\ O_2 \rightarrow 0.9\ CO + 0.1\ CO_2 + H_2,$$

a reaction in which the quantities of oxygen and water implemented are controlled so that the gasification is done without a contribution of external energy.

Assessment

The development of small-size reformers (30 to 50 kW) intended for the supply of fuel-cells for automotive vehicles is fairly recent and their performance (like their cost) is evolutionary and little known, but already very encouraging in their current state. Some American manufacturers (such as De Nora Fuel Cells SpA., Idatech, N-GHY (Fr), General Motors, Johnson Matthey, etc.) announce outputs higher than 80% in volumes with the same order of magnitude as the fuel cell itself. An output of 95% with methanol was even announced [COM] on a prototype whose response time was about 20 secs.

4.4.7.2. *Regulation and standardization*

It goes without saying in a fuel-cell generator including many components (compressor, pressurized enclosures, electrical equipment, etc.) and a hydrogen source, a regulation must be applied. However, the regulations and standardizations specific to fuel-cell systems and hydrogen are today available at European and international level.

Standardization

A technical committee, ISO[9] TC 197 "Hydrogen Technologies", was created in 1990 in order to work on standards for the devices concerning the production, transport, metering and use of hydrogen. France is represented there by AFNOR[10]. This committee consists of seven workgroups; Group 7 "Fundamental Safety

9 ISO: International Standard Organization.

10 AFNOR: Association Française de NORmalisation.

Prescriptions for Hydrogen Systems" is, in particular, in charge of working on a standard on the basic requirements for the safety of hydrogen systems.

Another technical committee, IEC TC 105 "Fuel-Cell Technologies", was created in 1996 in order to work on standards in the field of fuel-cell technologies, both for portable and stationary applications. France takes part in it via the UTE[11]. This committee consists of six workgroups.

On the other side of the Atlantic, organizations such as the National Hydrogen Association (NHA), the Canadian Hydrogen Association, the NFPA (National Fire Protection Association), the ANSI (American National Standards Institute) and the ASME (American Society for Mechanical Engineers) work on developing codes, standards and guides relating to the fuel cell.

In Germany, several organizations (BAM, TÜV, DIN) work on the development of texts applicable to hydrogen and the fuel cell and are, from this point of view, more advanced than in France, where the position would be now to await the development of the texts worked out by the ISO committees.

Regulation

In terms of regulation it is necessary to distinguish the various scopes of application (stationary, mobile and portable). For "mobile" applications the use of hydrogen requires the introduction of specific regulations, in order to guarantee a high security level. Thus, an initiative was taken at European level through the introduction of the "European Integrated Hydrogen Project" (EIHP), whose first phase, EIHP1, finished in 2000 and whose second phase, EIHP2, finished in 2004 [EIHP]. Its objective is to propose a

11 IEC: International Electrotechnical Commission.

European regulation for hydrogen and the distribution infrastructure, to replace the existing national regulations.

The first phase, EIHP1, made it possible to finalize two proposals of texts for the hydrogen vehicles: one for liquid hydrogen and the other for hydrogen gas. As a result of this work, the European Commission is soon to promote European projects aiming at having them recognized by national administrations.

4.4.8. *Conclusions*

Among the various types of fuel cell, it is the PEM fuel-cell whose application for cars is the most studied and the most probable in the next decades.

Having begun in the early 1990s, the development of this technology has known considerable progress, passing in less than ten years from the stage of laboratory curiosity to the stage of technology of the future, now considered as unavoidable, in the long run, by almost all car manufacturers.

Of the three obstacles which any technology must overcome (technical viability, environmental viability and economic viability), the first have now been overcome and the third should also be by 2020. Several manufacturers (DaimlerChrysler and Honda, primarily) began from the limited series of a few tens of vehicles, at the end of 2004, but most manufacturers do not consider true marketing before 2015-2020. Between now and then, a spectacular window is opening up in the eyes of the public, the California Fuel-Cell Partnership [CFCP] which as of the end of the year 2000 started to put in place a demonstration which will gradually bring together dozens of light vehicles and dozens of buses, of all origins, under regular operation at a real site.

Since 2002, the fuel cell intended for the propulsion has had a power from 50 to 75 kW according to the models, but is always hybridized with an electrochemical storage (battery Li-ion increasingly used, or supercapacity) which ensures an addition of power and recovers part of the braking energy in urban cycle.

Apart from this “drive” application, another scope of application has appeared in recent years in the automotive sector: it relates to the supply of electric on-board auxiliaries, increasingly greedy in terms of energy. Manufacturers like BMW and Renault have clearly stated their will to develop this type of fuel-cell generator, called APU. The power considered necessary by equipment suppliers to satisfy needs in the fifteen years to come is close to 5 kW. Two types of fuel cell could satisfy this new market: the PEM and the SOFC fuel-cell, both having been shown in operation on a vehicle by the manufacturer BMW. This application could see the light of day, in series, before the preceding one because of a less strong cost constraint (niche of the premium vehicle or heavy transport).

Concerning fuel, there are currently two different schools of thought: on-board hydrogen and on-board reformer, the first having had an unquestionable advantage at the beginning of the first decade of this century. The latter is further subdivided into two schools: reforming of a hydrocarbon (natural gas or synthetic gasoline) or a biomass alcohol (ethanol). Each one has its advantages and its disadvantages in the short and medium term, but all players agree that in the long run, on-board hydrogen will be the best solution, hence the preference of some to go immediately to this solution. The major question is knowing when this last phase will begin: ten, twenty or thirty years? According to the strategy retained, the implications on the possible choices of reforming are not insignificant, all the more so as even if hydrogen is retained as a fuel stored on-board, it is probable that part of this gas will come from a reforming in a station!

Lastly, considering the choice of hydrogen in the long term, several solutions appear to store it on-board: both solutions realizable today (storage under pressure or on hydrides are most probable) and those possible in the long run (storage in nanostructures).

4.5. Bibliography

[ALL 99] ALLEAU T., SERRE-COMBE P., *Le stockage de l'hydrogène*, symposium *JEPAC'99 Piles à Combustible*, Paris, 25-26 February 1999.

[BAU 87] BAUSSIÈRE Robert, LABRIQUE Francis, SÉGUIER Guy, *Les convertisseurs de l'électronique de puissance*, vol. 3, *La conversion continu-continu*, Tec et Doc, Paris, 1987.

[BON 98] BONAL J., SÉGUIER Guy, *Les Entraînements électriques à vitesse variable*, tome 2, Tec et Doc, Paris, 1998.

[BOS 82] BOSE B. K., *Power Electronics and AC Drives*, Prentice Hall, Englewood Cliffs, New Jersey, 1982.

[CFCP] *California Fuel-Cell Partnership*: http://www.fuelcell partnership.org

[CHA 94] CHAHINE R. *et al.*, "Low pressure adsorption storage of hydrogen", *Int. J. Hydrogen Energy*, 19, 161, 1994.

[COM] Compact Methanol Reformer Test, Design, Construction and Operation of a 25 kW Unit, *Joule III Contract*, JOE3-CT-950038

[EIHP] EIHP, site http://www.eihp.org

[GRETh] these data are available in particular from GRETh (Groupement Recherche Echangeurs Thermiques) a part of the CEA-Grenoble.

[HYD] Site http://hydpark.ca.sandia.gov

[LEO 90] LEONHARD W., *Control of Electrical Drives*, second edition, Springer Verlag, Berlin, 1990.

[NAN] *The Nanotube Site*, http://www.pa.msu.edu/cmp/csc/nanotube.html

[ROD 96] RODRIGUEZ N., BAKER T., *Graphite Nanofibers: the Answer to Hydrogen Storage,* Material Research Society, Fall Meeting, Boston, 1996.

[SEG 89] SÉGUIER Guy, LABRIQUE Francis, *Les Convertisseurs de l'électronique de puissance – la conversion continu-alternatif*, Tec et Doc, Paris,1989.

[SHO 00] SCHOTT P., POIROT J.-P., BAURENS P., *Modélisation et simulation de la source d'énergie à pile à combustible du véhicule Hydro-Gen*, *Pile à combustible et interfaces pour les transports*, conference, Belfort, France, 9-10 November 2000.

Chapter 5

Prospects and Evolutions of Electric-Powered Vehicles: What Technologies by 2015?

5.1. Mobility

Transport activity is characterized both by its extreme diversity and its intersection with all other spheres of economic and social life, on all scales of the territory.

In a century, technological evolutions have made phenomenally great strides in mobility, which has largely contributed to reorganizing our societies. It has supported economic, social and cultural development.

The main tendencies of this mobility are the following.

The era of megalopolises

By supporting the movement of goods and people, mobility has contributed to the creation of major urban centers. In 2005, half of the world's population will live in urban areas,

---

Chapter written by Joseph BERETTA.

compared to 30% in 1950. The phenomenon is particularly strong in developing countries. In 2000, 15 of the 19 megalopolises were in developing countries. In 2015, 18 of the 23 megalopolises will be located there.

Sustainable mobility

We speak about this more and more; it is the merging of mobility and sustainable development, or how to support an equitable mobility for the greatest number, limiting its environmental effects at an acceptable cost. It is by necessity a complex and long-term approach, requiring thinking as to the complementary action of modes of transport and the answers to the real needs of the population.

Further, but not much further

Because of the increase in modern societies' urbanization, the distances covered increase each day. However, they have progressed more quickly than the time devoted to transport, proof that, contrary to generally accepted ideas, traffic conditions are improving. The time-savings achieved by the improvement to mobility are invested in distance.

Advantage with cars

All the studies show that the car is by far the preferred means of transport for Europeans. This unanimity is confirmed by a recent study by "Thalys international", in which 64% of Europeans judge the car practical and 42% appreciate its driving comfort. It is fairly understandable, considering that a private car makes it possible to go where we want when we want, carrying out complex journeys, at several stages, which no collective transport makes it possible to do.

Collateral effects

In recent decades, the rise in mobility was accompanied by a certain number of collateral effects, like polluting emissions and noise. The reaction of manufacturers made it possible to strongly reduce them. Recent vehicles, for example, pollute a hundred times' less than vehicles fifteen years ago and are approximately half as noisy on the outside. And contrary to a widely held idea, road traffic is improving, in particular for journey times, at least in developed countries. There remains the question of CO_2: European and Asian manufacturers have committed to reducing their cars' specific emissions by the end of the decade. In Europe, the projects in progress will lead to a reduction of the CO_2 emissions of road transport as of 2012 (CO_2 regulation = 130g CO_2).

New horizons of automotive mobility

New needs as regards mobility are pushing manufacturers to promote new practices with regard to the car.

Innovating services with a better integration of the private car in the public transport network is one example of that.

20 years ago, the use of the car was reserved for a few essential journeys: work, weekend, and holidays. However, with the evolution of lifestyles, the development of part time, the reduction of working time or the advent of a more active senior generation, reasons for traveling have multiplied, and needs have changed.

More individualistic consumers want to keep control of their travel as much as possible, and maximize the value of the time spent in transport. Queuing up at a counter or remaining stuck in traffic is becoming increasingly less tolerated. Everyone wants to move "without obstacles" or waste of time.

Public transport has tried to adapt to this evolution by introducing new services into their scope: information on journey times and delays, automated reservation machines, convenience stores in stations or subway stations, left luggage services, etc. The car has also adapted to this need for intelligent mobility. Navigation systems have been conceived to facilitate journeys and to make them more effective.

But the phenomenon should quickly develop and give rise to new ways of consuming the car. The car has a considerable potential for adaptability and diversification; we simply need to regard it not simply as an object but as a pallet of solutions. The traditional “car as a product”, that we buy and that we possess can very well double as a “car as a service” which we borrow and consume.

The concept of a self-service automotive system is one of the answers to the problem of mobility.

By proposing the self-service use of the electric car, Liselec widens the range of urban mobility possibilities. Complementary to other modes of displacement, it offers its subscribed customers a service which merges the flexibility of private car use and the advantage of public transport. Liselec is made up: of a completely silent vehicle that does not produce any urban pollution, and of stations (car-park areas) and recharging points spread throughout the city. This service has been in operation since 1999 in the city of La Rochelle.

A service such as Liselec is a first stage necessary to arrive at the concept of an urban vehicle which answers the needs of users and the difficulties of city traffic jams. Everyone wishes for transport available 24-hours-a-day which enables them to travel from door-to-door, which neither buses or subways offer. But if each individual takes their vehicle to go into the town center, we run up against the number one problem of all big cities: the congestion due to the lack of traffic space resources and parking areas. The solution thus lies in a self-service system of shared vehicles. Liselec and Praxitèle made it possible to test this principle (experiment at St-Quentin-en-Yvelines, using self-service electric Renault Clio vehicles). It is necessary to be able to constantly ensure availability in any place. The handicap of self-service lies in the stations: users must go to them to pick up and return their vehicle, and they are not necessarily close to their starting point or destination. Then it is necessary to rebalance the fleet from one station to another.

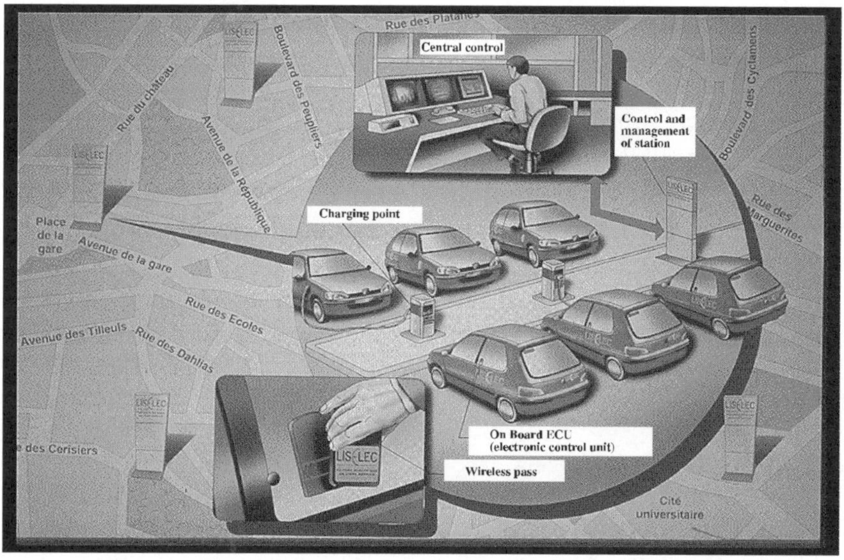

The stages which remain to be overcome are not technological. If we want to circulate non-polluting vehicles in an urban site, it will be necessary to limit the access of conventional cars, which affects the fundamental freedoms of our civilizations. Lastly, certain customers remain very much attached to the pleasure of driving. But where is the pleasure in driving bumper-to-bumper every morning to go to work?

In certain European countries, like Switzerland, Germany or Holland, an alternative solution to conventional automotive services has thus developed over the last ten years: car-sharing. The principle: a group of individuals come together in a company or a co-operative to buy vehicles, which it then places at the disposal of its members for their travel, with favorable economic conditions due to the sharing of the fixed costs. The main advantage of the system: the removal of the constraint and associated with owning a vehicle. The flexibility of car-sharing makes it possible to borrow a vehicle for one hour, which traditional renting does not permit. We can speak about individual means of public transport. This car approach is extremely complementary to public transport.

The advantage of the car, it is that it makes it possible to manage small flows, to meet needs for interstitial or diffuse mobility. If the inter-modality exists between the collective modes (to pass from the train to the bus), it now remains to create a inter-modality of the car, while making it possible to the users to pass from their personal car to a hiring for the weekend or to car-sharing in order to move during the evening or to make their shopping trips.

5.2. New technologies

As it is generally the case, it is not an evolution or an industrial innovation which makes it possible to take a big important step in the changes of the technologies implemented, but rather a coupling and a mixing of several tendencies which will build the future trend.

Thus, to pass from the thermal car to the electrified car, several major tendencies all impacting in the same direction will allow convergence towards a credible scenario:

– the constant evolution of the electrification of functions (thanks, amongst others, to the evolution of the components) will in the long term involve switching from the 12V to the 42V network, making it possible to provide an electric output of about 5 kW and the implementation of alterno-starter concepts;

– the taking into account of CAFE 2012 (CO_2 regulation) objectives involves a global concern for the optimal use of energy;

– the increasing introduction of multiplexing will introduce more and more electronics into distribution and commutation functions.

The first consequence: the electromagnetic relay will be replaced by static switches (because the cost of silicon at constant commutated is lower than the relay in 42V), which will allow the introduction of additional functions such as diagnosis, load unbalancing and degraded modes, at reduced cost.

The development and the use of these 42V power components will make it possible to standardize the power electronics that ensure its introduction into other systems. Power electronics will finally answer automotive constraints; its generalization with electric drive systems will be facilitated.

The ultimate target will be the “electrified vehicle”, i.e. using a thermal engine, assisted by an electrical motor. The power network will be 42V implementing solenoid valves, electric steering, electric air-conditioning, etc., using, of course, batteries of new technologies.

The domino effect can continue and allow the possibility of designing electric vehicles using electric parts coming from the major flow and presenting a balance between service and cost.

The electrification of vehicles is underway and will develop in years to come. But switching to 42V introduces more economic than technical problems. It is thus natural, initially, to optimize today's systems by managing electrical energy as well as possible (unbalancing, proportional control of the actuators recovery of energy). This makes it possible to push back the time limit to after 2010.

Even when even the functions are used as advised, the electric output of the power networks will increase because certain large-scale loads will become necessary (electric steering, solenoid valves, temperature setting of catalyst system, etc.). This electrification will only complicate the problems, already serious today, of EMC (Electromagnetic Compatibility).

5.2.1. *Electric motors*

The major principles of electric motors (synchronous, asynchronous, DC current, switch reluctance, etc.) have been mastered and do not evolve. The innovations relate to basic interconnections, new designs and improvements in materials.

There is no such thing as a good or bad electric machine, but motors adapted to each application.

Thus, the DC-current motor, which does not have a great power density, suffers from a certain fragility and an additional cost due to its collector and it is still used for the small actuators (windshield wiper, ventilation, seat and door controls, etc.) because it is very well adapted to the low-voltage automotive electrical network (DC 12V). It does not require electronics for its control and its supremacy will only be called into question with the introduction of the 42V and the need for fine control.

For power applications like traction, the tendency is to use alternating-current motors. Among this family, the synchronous magnet motor seems to be favored. The concerns raised about the availability and cost of permanent magnets did not prevent the virtually exclusive presence of synchronous magnet motors on all the hybrid vehicles presented recently. Its very good power density explains why it is selected in hybrid vehicles when the problems of integration are significant.

The basic interconnections of motors can give interesting results.

The magnetic materials and the insulating materials have to work either in a hot environment or in a moderate environment but with power densities such that the iron losses of the components would raise their temperature up to 200°C. The constraints imposed by the rise in power per unit of volume led to the development of materials likely to work at more than 200°C.

The design (internal geometry) defines the electric motor's performance. The number of possibilities is virtually infinite. To meet a need, suppliers are satisfied with breaking down an existing geometry, performing homotheties, for example. Even if the solution obtained works well and meets the need, this conceals the existence of a more optimal design in terms of efficiency, volume and mass.

With the need for electric motors being very varied in the automotive field, each motor technology will be able to find its scope of application there. Efficiency and compactness will remain the major advantages of synchronous magnet machines. Asynchronous and switch-reluctance motors have the significant advantage of a no-auxiliary supply (thus without losses) even if they are not mechanically driven by the other systems of the vehicle. The "solid cast" rotors of the switch-reluctance motor allow them easier access to high speeds.

5.2.2. *Electronic power systems*

The reduction of the size and the weight of energy transformation devices has become a major issue for power electronics. The tendency to work at increasingly higher temperatures seems to be confirmed. The structures of the components move very little. The appearance of the IGBT (Insulated Gate Bipolar Transistor) in 1981 was a major advance. The IGBT is power component integrating a MOS and a bipolar transistor into the same structure. Their association makes it possible to combine the advantages of these two elements while minimizing their disadvantages as much as possible. Since its creation, the enormous technological advances that have occurred have enabled the IGBT to acquire very good reliability. It is a component which is thus used in numerous, varied applications. The IGBT has become, in less than ten years, a recognized component. With the MOSFET reserved more for low voltage, they will share the car applications.

In power electronics, we would be right to think that for a certain number of applications, other materials like silicon can be used, and particularly large-gap materials such as SiC (silicon carbon), diamond, etc.

These materials find their justification only when we can push the ultimate limits: voltage, temperature, frequency. Studies concerning an environment which is specific for them, and which allows them an optimal operation must thus be developed.

The current interest for ceramic materials is mainly associated with the possibilities of obtaining often remarkable electric properties (high permittivities, correct field of rupture, good thermal conductivity, etc.), on the one hand, and in addition with the possibility of using them in frequency or temperature ranges from which polymers are generally excluded.

Contrary to the structures (IGBT, MOSFET, Thyristor, GTO, MCT), the packaging of the component is evolving. Cars represent an interesting market for component manufacturers, if the production volumes are sufficient. Some manufacturers have already proposed dedicated components. Thermics is more important for power electronics, and packaging is moving in this direction, taking into account reliability.

5.2.3. *Electric energy sources*

Electrochemical storage systems

First of all, an important point to highlight is that in the field of electrochemistry, the evolutions are always slow because of complexity of the chemical systems implemented.

To guarantee the success of a technology in car applications, it is essential that they answer the three criteria: cost, safety and performance.

Various technologies which will be present in 2020:

– lead acid (conventional (open) and sealed);

– Ni-MH;

– lithium (liquid electrolyte Li-ion, Li-ion lamitate, lithium polymer gel electrolyte, Lithium polymer solid electrolyte).

Lithium: main candidate for applications: Mild Hybrid, Full Hybrid and EV if it confirms its potential for cost reduction.

Lithium ion: it should be broken down into various technologies:

The liquid electrolyte Li-ion: operating life and performance are now validated. The problems involving safety are on the way to being solved;

The Li-ion polymer laminate, from companies specialized in portable applications and which are now taking an interest in the transport market.

The solid electrolyte Li-polymer is today still at the study stage. The research projections on the polymer membrane will make it possible to have batteries working at room temperature. Today the operating temperature of these systems is 80°C. It will be necessary to have a heating system for low temperatures. The process, connected to the plastic films industry, should lead to a reduction in the cost of the cells. The cost of the packaging and thermal management will be higher than for the Li-ion, however.

In addition to the effects of volumes and the arrival of Asian players on the market, reductions are expected in the cost of materials. Great hopes are placed in the cathode containing iron phosphate (in place of oxides Ni, Co or Mn).

Ni-MH: as in mobile applications, it will see its market shares eaten away at by Lithium.

Lead acid: owing to its low costs, lead acid is likely to remain the only candidate economically viable for implementations of 42 V starting and power Net, 42 V with stop/start function, 42 V for soft hybrid.

Let us recall that there exist two types of battery technology and two types of design element:

– tight and open technology: tight batteries have a greater capacity to cycle compared to conventional batteries (open lead). In addition, tight batteries have the advantage of being able to be placed inside the cabin without the risk of an acid leak or gaseous emission;

– flat and orbital (spiral) electrode design: electrodes of spiral design offer better power than that of the conventional prismatic design.

Consequently, it is the fact that they must present cycled capacities to fulfill the functions of hybridization (stop-and-start hybrid application) which will impose the passage to the tight method. Indeed, we can very well imagine 42V open lead-acid batteries to provide a Power Net function without hybridization.

The switch from tight technology to the spiral design will make it possible to obtain more powerful batteries to fulfill the functions of hybridization for traffic congestion hybrid applications.

Consequently, by 2020 the following will coexist:

– the conventional lead-acid which will undoubtedly be present for stop-and-start applications and 12V- and 42V-power network systems, orbital tight lead-acid for 42V applications with stop-and-start and stop-and-go (mild hybrid).

5.3. New cars

On the path leading to hybrids, there is no time to lose

The democratization of hybrids is announced in the next ten years. Associating electrical energy and thermal energy, hybrid vehicles are being studied by all the major manufacturers and are already on the road for some of them. The objective: to push the reduction in consumption and polluting emissions even further.

In over a century, the thermal engine has widely demonstrated its effectiveness. But it is reaching a plateau today, in terms of consumption and polluting emissions, which it will become increasingly difficult to improve upon. As for the electrical drive train, which is perfect for the environment (the electricity production system aside), it remains handicapped by a limited range and a high cost. Consequently, why not associate both to add together their strong points, and reduce their weak points? It is the principle of hybrids, vehicles which use these two energy sources. They are widely talked about, even if the worldwide automotive market still counts few of them: the Toyota Prius, Estima and Crown and the Honda Insight and Civic and Citroën C3 stop and start. There are, at the beginning, two major ways of approaching this coupling.

The first, nearest to current cars, consists of adding an electric system to a conventional power unit (motor, electronics, battery of variable power). We then speak of parallel hybrids.

The second is to make use of a thermal engine as a generator to supply an electrical motor able to drive the wheels. This is the world of series hybrids.

The world of series hybrids

Mechanical energy transformation into electrical energy requires large electric motors and involves output losses throughout the drive train. The assessment of the "series" was relatively poor, until engineers had the idea of taking part of the mechanical power to transmit it directly to the wheels. We then speak of hybrids with power spilt, a recent branch, to which the Prius belongs. Thanks to this clever trick, it becomes possible to halve size of the electric motor compared to a conventional "series", with identical services. Nevertheless, in spite of this promising evolution, the "series" continues to suffer from another handicap: the transfer of mechanical energy into electrical energy is done in the range of the "full hybrids". Yet, on this level, we are at "high" voltage, which requires, at the same time, protections (for the protected electrical network), non-standard components and extremely expensive batteries, not easily compatible with the cost constraints of cars.

Parallel hybrid, the most realistic method

The parallel hybrid seems, in the medium-term, to be the most realistic method and most practicable in Europe. The gains hoped for with a "parallel" vary according to the level of voltage and power which the manufacturer chooses to establish on the vehicle. Logically, the higher this level, the more the possibilities are numerous, but the costs soon add up. By preserving the voltage of current vehicles, that is to say 14 volts, and by establishing an alterno-starter with a power of 2 kW, we can make a "stop & start", i.e. cut the thermal engine during shutdown and start it again quickly, in order to stop consumption during non-productive phases, an opportunity that is of particular interest for town driving. The Citroën C3 stop & start is the first commercialized car equipped with the stop & start system. Maximum estimated gain: around 10% of consumption in a dense urban cycle. By making the voltage climb to 42 volts and the power in a range from 5 to 15 kW, we reach the world of the mild hybrid, which makes it possible to go further by assisting the thermal engine in the more penalizing consumption phases (with low torque, in traffic jump, for example). In addition to the stop & start, the vehicle opens the way to the "brake & boost", where the electric system comes to help the thermal engine during acceleration and recover braking energy to supply the batteries during deceleration phases. If the power is carried at around 10 to 15 kW, it becomes possible to move on to an electric drive: the vehicle is able to run in all-electric mode in the starting phases or during low-load driving, without using expensive batteries known as "of energy", necessary to the full hybrid. Ultimately we obtain greater driving comfort, more torque at low speed without a response time and a potential gain in consumption of 15 to 20%. By increasing the voltage (150 to 500 volts) and the powers (50 to 100 kW) more extensively, we then enter into the world of the full hybrid, which, in the longer term, could cover all that we can imagine of a hybrid vehicle: pure electric drive or simple electric assistance of the thermal engine, recovery of energy during braking, stop & start. In addition to the fuel savings achieved on the kilometers travelled electrically, such a system could lead to gains in consumption of about 30 to 40% in hybrid mode, compared to a thermal vehicle.

Nevertheless, the cost/service balance for the full hybrid seems difficult to obtain on the European markets. The industrial stage of the "mild hybrid" is possible only with fields of convergence with the conventional vehicle (engine, gearbox, 42V network), in order to reduce the specific costs.

The first field of convergence: "downsizing", a basic tendency which consists of obtaining with a small engine the performance of an engine of higher cubic capacity, by exploiting all the registers of optimization – high-pressure injection, common rail, turbocharger, etc. The disadvantage is that these small engines are subjected to strong stresses at low speed. The idea is to help them in these operating ranges with an electric motor, to gain in drivability and to reduce fuel consumption by 10%.

The second field of convergence: the expected switch of the power network of all new vehicles from 14 to 42 volts, which will support hybridization. "However, the industrialization of a mild hybrid is possible only if the additional cost at purchase can be compensated by a gain in consumption and, as much as possible, an improvement of the drivability". But, several barriers (electric motors, battery and power electronics) still remain to be removed on hybrid devices.

Technological barriers

The first technological barrier is the electrotechnical components: they need to gain in compactness (integration), efficiency (consumption) and cost. The second barrier is the batteries; lead-acid technology will need to give way to that of super caps or lithium-ions, more compact and lighter batteries with equal power, but also more expensive. Yet for all this, removing these two barriers is not enough.

The key to hybrid success is the optimal sizing of all of the parts and good management of the energy: a leader (an electronic supervisor) is essential to determine the optimal operating mode between thermal engine, electric motor and auxiliaries. It is necessary in order to limit the costs and avoid the pitfalls of oversizing.

Lastly, manufacturers will have to find external partners to develop and produce in series the electric motors, their control electronics and batteries.

On the way to a fuel-cell vehicle, it is time to accelerate slowly

Fuel-cell vehicles are announced for the second half of the millennium, when oil will be rare and when fuel-cells will have achieved their cost and reliability objectives.

PEMFC technology

The operating temperatures of the membranes should increase to be around 120°C. The operating life for stationary applications would be satisfactory and the same for drive applications.

The bipolar plates are likely to evolve towards metallic solutions which allow a reduction in overall dimension. Their operating life should strongly progress.

Concerning the stack, the main hard point will remain its cost which is likely to remain beyond the automotive objectives for a drive-train. This is mainly related to the material cost. Its operation on a broad temperature range should be controlled.

Other fuel-cell technologies.

SOFC technology is likely to develop for APU applications with thermal engines in order to satisfy future electricity consumption. It is not likely to experience a drive-train application.

DMFC (Direct Methanol) technology is likely to develop for mobile applications (telephone but especially computer). It is not likely to be retained for vehicles, but it remains a method to be monitored. Indeed, in spite of a lesser performance, it makes it possible to use a liquid fuel without a reformer.

Alkaline fuel-cell technology will still be at the R&D stage.

Reformer system

Manufacturers are giving up on methanol.

The technology of the gasoline reformer presents a certain number of hard points which are likely to remain crippling for an automotive application, like the thermal management of the various stages of the reformer and the complete system's overall dimensions, and last but not least, start-up times.

In addition, the cost of such systems is likely to remain high.

Storage of hydrogen

Compressed storage at 700 bars is likely to be the most widespread technology for automotive application. An international regulation will exist. The cost of this type of tank could remain a difficult point. Moreover, this technology is dependant on the accident risk inherent to gas storage.

The cryogenic storage method is not likely to be developed for cars but could be continued for fleets of buses.

Methods presenting larger safety will continue to be worked on.

The technology of solid storage in hydrides will be optimized for an automotive application but is not likely to reach the industrialization stage. With regard to safety, while the on-board hydrogen pressure is less, this technology does not completely answer safety concerns because there always remains residual hydrogen at 10 or 30 bars in the tank.

The potential for storage in nanostructures of the nanotube type will be completely evaluated.

Chemical storage technology should be viable and implementable in a car.

It would be a good alternative in the event of a security issue presented by compressed storage. On the other hand, it requires a network for the recovery and recycling of borate, which will penalize it.

Public authorities

The US are massively increasing their efforts in the fields of the fuel-cell and hydrogen, their main motivation being energy independence. In 2015, the US should have taken decisions and set up the actions necessary for the transition towards a hydrogen economy.

The actions of the federal government and certain states will be decisive in the installation of a hydrogen distribution network and diversified means of production making it possible to ensure their energy independence.

Japan is also increasing its efforts in these fields. Its level will remain close to that of the US.

Today, Europe is lagging behind, both in terms of the budgets authorized for R & D and in the progress made with programs. Between now and 2015, the EC will continue its efforts through the hydrogen platform because it considers this to be a strategic domain.

The hydrogen network

Ten European capitals are installing hydrogen service stations for captive fleets of buses. These stations make it possible to test various technologies of distribution (hydrogen in gas or liquid form for the moment) and provisioning (reforming, electrolysis of water, etc.).

In the US and Japan similar experiments have already been conducted to validate hydrogen distribution technologies. They fuel trucks, buses and passenger cars.

In 2015, taking into account the costs of fuel-cell vehicles, the problem of implementing a hydrogen distribution network for the general public remains significant. A distribution network for captive fleets will exist and will be managed by petroleum companies and energy suppliers.

The installation of a network for the public will require strong public/private support and will remain a difficulty taking into account the fuel-cell market, which is still small.

The market:

In 2015, the automotive market of fuel-cell vehicles will be made up exclusively of captive fleets. In the US and Japan, these fleets will, however, reach volumes which could be significant (several thousand per year). In Europe, this market will exist but will be less developed.

"One thing is sure: we can no longer consider the future only from the angle of fossil energy use by an internal combustion engine. It is a solution which has done great service and will continue to do so, but not in such an absolute way as during the last century. However, it would be just as illusory to think there exists an alternative miracle solution, both in terms of energy and the environment. The uses of the car and ways of being mobile are increasingly numerous. It is thus logical to think that for different uses and different places, there are different corresponding technological answers. Electric, hybrid in various forms, alternative fuels, fuel-cell vehicles:

The future is one of diversity."

Joseph BERETTA.

Automobile Glossary

A

adaptive automatic transmission
Some transmissions can "learn" and adapt to the driver's style, altering shift points and other transmission functions to produce the most efficient operation.

adaptive cruise control
One step beyond cruise control, adaptive cruise control uses a radar sensor mounted in the front of the vehicle to maintain a safe cruising distance between your vehicle and the vehicle in front of you. Most systems allow you to set the interval (distance between vehicles) and will then accelerate and brake (within reason) to maintain that interval. In many cases, the systems can take into account hilly terrain and passing situations. All have the ability to act like regular cruise control, and they still require the driver to steer the vehicle normally. Some car manufacturers refer to it as "intelligent" cruise control.

ABS (anti-lock brakes)
Helps maintain control in emergency stops. Particularly useful on slick pavement. While the driver applies steady pressure to the brake pedal, the system automatically

"pumps" the brakes many times per second to prevent wheel lock-up.

ACEA

Association des Constructeurs Européens d'Automobiles (European Automobile *Manufacturers Association)* (14 members), whose head office is located in Brussels. Its members are, in addition to PSA Peugeot Citroën, BMW, DAF, FIAT Auto, FORD Europe, GM Europe, MAN, Daimler Chrysler, Porsche, Renault, Rolls-Royce, Scania, Volvo, VW and Toyota.

In 1998, the members of the ACEA carried out 40% of the worldwide automobile production, that is to say, nearly 22 million vehicles.

Address:
Association des Constructeurs Européens d'Automobiles
Rue du Noyer, 211
B-1000 BRUSSELS - Belgium

ACV

An abbreviation for air-control valve.

aerodynamic force

A force created aerodynamically, such as by an air foil.

aftermarket

Equipment sold to consumers after the vehicle has been manufactured. Aftermarket equipment and parts are sold by catalog, dealers, independent garages and parts houses.

air

The combination of gases that make up the Earth's atmosphere: nitrogen (76–78%), oxygen (18–21%), and small amounts of carbon dioxide, argon, and other gases. When air is drawn into an engine, the oxygen combines with the fuel during combustion, producing carbon dioxide and water vapor.

air bag
Passive restraint with an inflatable air-bag located in the steering wheel in front of the driver and in the dash in front of the right-hand front-seat passenger.

An inflatable bladder used in place of a spring in an air-suspension system. Also see air lift.

air conditioning (A/C)
The process of adjusting and regulating, by heating or refrigerating, the quality, quantity, temperature, humidity, and circulation of air in a space or enclosure; to condition the air.

all-wheel drive (AWD)
A vehicle drive-train with a center differential having all four wheels under power at all times.

alternating current (ac or AC)
The type of electrical current actually produced in an alternator.

alternator
An electricity-generating device that converts mechanical energy to electrical energy in the form of alternating current. Diodes rectify the alternating current into direct current.

amp
An abbreviation for ampere.

ampere (A)
A unit of measure for current.

amp/hour
Amperes per hour; a standard measure for a current flow rate.

anode
The positively charged electrode in an electrolytic cell toward which current flows.

atmospheric ozone
As ultraviolet (UV) rays from the sun reach Earth, they are combined with smog and other pollutants to produce atmospheric ozone. Atmospheric ozone, unlike stratospheric ozone, is considered harmful. Whenever possible it is to be avoided.

alterno-starter
New architecture which includes an electric motor managed electronically, intercalated between the engine and the gearbox, replacing the starter, the alternator and the flywheel. In addition to the battery and silent and instantaneous start-up, it is able to constantly provide an auxiliary power of approximately 5 kW. A "stop and go" function makes it possible to save up to fuel in urban driving. It also makes it possible to smooth the acyclisms of the main engine and facilitate hill starts. Lastly, it makes it possible to have, inside the cabin, 220 volts to connect common electrical appliances.

ASR
Anti-Skid Regulation. Electronic device making it possible to reduce or even remove the skidding of the drive wheels by acting on the brakes, and possibly on engine management.

B

battery
A device for storing energy in chemical form so it can be released as electricity.

battery capacity
The energy output of a battery measured in amp/hours.

battery cell
The part of a battery made from two dissimilar metals and an acid solution. A cell stores chemical energy for use later as electrical energy.

battery charge
The restoration of chemical energy to a battery by supplying a measured flow of electrical current to it for a specified time.

battery charger
An electrical device that is used for restoring a battery to its original state of charge by passing a current through the battery in a direction opposite to the discharge current flow.

big three
The three major United States automotive manufacturers, Chrysler, Ford, and General Motors.

body
The assembly of sheet-metal sections and other parts that provide the enclosure of a vehicle.

buggy
A small, lightweight, off-road vehicle.

biofuels
Fuels either of plant origin or resulting from a fossil mixture of fuel and vegetable additive (e.g. Diester, 30% methyl plant oil/vegetable oil/oil esters, 70% diesel).

C

cab
The driver/passenger compartment of a truck or off-road vehicle. Or short for taxicab.

California Air Resources Board (CARB)
A California agency responsible for regulations intended to reduce air pollution, especially that created by motor vehicles.

canister
A container filled with charcoal in an evaporative control system used to filter and trap fuel vapors.

carbon monoxide (CO)
An odorless gas composed of carbon (C) and hydrogen (H), formed by the incomplete combustion of any fuel containing carbon. This major air pollutant is potentially lethal if inhaled, even in small amounts.

carbonyl chlorofluoride (COC1F)
A toxic by-product of Refrigerant-12, and other CFCs; toxic if allowed to come into contact with an open flame or heated metal.

carbonyl fluoride (COF2)
A toxic by-product of Refrigerant-12, and other CFCs; toxic if allowed to come into contact with an open flame or heated metal.

catalyst
A lead-sensitive substance, such as platinum, palladium or rhodium, that accelerates or enhances a chemical reaction without being changed itself. When used in a catalytic converter, it can reduce the level of harmful pollutants in the exhaust.

catalytic converter
An automotive exhaust-system component, made of stainless steel, containing a catalyst to reduce nitrogen (NOX) and/or hydrocarbon (HC) oxides, and carbon monoxide (CO), in tailpipe emissions.

carbonyl chlorofluoride (COC1F)
A toxic by-product of Refrigerant-12, and other CFCs; toxic if allowed to come into contact with an open flame or heated metal.

center of gravity (CG)
The exact point around which an object, such as a vehicle, is perfectly balanced in every direction.

center of wheelbase
The exact point midway between the front and rear wheels of a vehicle.

central processing unit (CPU)
The component of a computer system with the circuitry to control the interpretation and execution of instructions.

centrifugal force
The outward force, away from the center (axis) of rotation, acting on a revolving object, increasing as a square of the speed.

cetane
A primary reference fuel (C16H34) for describing the ignition quality of diesel fuel.

CFC
An abbreviation for the chlorofluorocarbon family of refrigerants.

CFI
An abbreviation for continuous fuel injection. Or An abbreviation for central fuel injection.

chassis
The lower structure of a vehicle to which the body and running gear are attached.

CIS
An abbreviation for continuous injection system.

club coupe
A two-door, four- or five-passenger vehicle with limited rear seating.

cluster gear
A set of three or four gears on a common shaft in a manual transmission.

clutch
A device for connecting and disconnecting the power flow between the engine and standard transmissions, used during starting, shifting, and stopping.

CNG
An abbreviation for compressed natural gas.

CO
A symbol for carbon monoxide.

coach

1. A bus.

2. A motor home.

3. An enclosed, two-door sedan of yesteryear.

4. A large, four-wheeled carriage; the forerunner of the modern vehicle.

coefficient of drag (Cx)
A measure of the air resistance of a moving vehicle; a measure of how much air is moved as the vehicle moves from one point to another.

coefficient of friction
The measure of the resistance of one surface moving against another.

combustion emission control (CEC)
An exhaust emission-control system that combines a transmission-controlled spark system and a deceleration throttle-position device.

compression ratio (CR)
A measurement of how much the air/fuel mixture is compressed inside an engine cylinder. If compressed to 1/10 of its original volume, the compression ratio is 10 to 1.

continuous fuel injection (CFI)
A type of fuel-injection system that continuously sprays a stream of fuel into the engine.

continuous variable transmission (CTV)
A stepless transmission that uses a sheave clutch to transmit engine torque.

convertible
An automobile having a retractable or removable roof.

cruise control
A control system that allows the vehicle to maintain a preset speed, though the driver's foot is off the accelerator.

common-rail
Input common hopper, maintained at very high pressure, which allows direct injection of the fuel into the cylinder. Each component of the system is controlled by a computer and allows an ideal injection volume, according to the load.

D

dashboard
The section immediately behind the windshield that houses the instruments, accessory controls, and glove box.

daylight
To be ahead of a competitor by more than a car length in side-by-side racing, so daylight is visible between the two vehicles.

DC
An abbreviation for direct current.

detonation
A phenomenon of internal combustion where the compressed air/fuel charge explodes violently instead of burning smoothly, usually due to the creation of a second flame in the front of the combustion chamber, away from the spark plug.

defroster
The part of the heater system designed to clear heavy frost or light ice from the inside or outside of the windshield.

diagnosis
A standard procedure that is followed to locate and identify the cause of a malfunction.

diesel cycle
An engine operating cycle where air is compressed and fuel is injected at the end of the compression stroke, causing ignition.

diesel engine
A compression-ignition engine.

differential
The section of the rear-axle assembly that provides three functions: it allows the wheels to revolve at different speeds during turns, provides the final gear reduction, and changes the angle of drive 90 degrees.

dimmer switch
A switch that allows the driver to select either high- or low-beam operation of the headlights, and to switch between the two.

DIN
An abbreviation for Deutsches Institut für Normung.

DOHC
An abbreviation for dual overhead camshaft.

drive line
Assembly of various parts such as the driveshaft, universal joints, and connecting yokes that transmit torque from the transmission to the differential.

drive train
All of the components required to deliver engine power to the road surface.

E

ECM
An abbreviation for electronic control module.

ECU
An abbreviation for electronic control unit.

E85
A fuel blend of 85% ethanol and 15% gasoline.

EGR
An abbreviation for exhaust gas recirculation.

electrolyte
A substance in which the conduction of electricity is accompanied by chemical action.

electromagnet
A device consisting of a ferrous-metal core and a coil that produces appreciable magnetic effects when an electric current exists in the coil.

electronic brake control module (EBCM)
A system having a monitor at each wheel to sense conditions and feed an electrical impulse into an on-board computer to reduce sideways skidding during rapid braking action.

emissions
Unwanted, harmful chemicals and chemical compounds that are released into the atmosphere from a vehicle, especially from the tailpipe, crankcase, and fuel tank, including unburned hydrocarbons, carbon monoxide, nitrogen oxides, particulates and sulfur.

engine torque
A rotating, twisting action developed by the engine that is measured in pounds, feet or kilowatt hours.

ethanol
A form of alcohol, C_2H_5OH, found in alcoholic beverages and also used as an additive in gasoline to produce gasohol.

evaporation control system (ECS)
A system that prevents fuel vapors from escaping to the atmosphere while the engine is off.

F

FISA
An abbreviation for *Federation Internationale du Sport Automobile* [International Motorsport Federation].

flywheel

1. A heavy metal wheel with starter ring gear that is mounted at the rear of the crankshaft. It absorbs energy on power stroke(s), returns energy on other stroke(s), and transfers power to the clutch or torque converter.

2. Front-most part of a clutch assembly that is bolted to the engine crankshaft with a rear surface to provide a smooth friction area for the disc-front facing to contact during clutch engagement.

fossil fuels
Fuels formed underground from animal and plant matter by chemical and physical change, such as coal, petroleum, and natural gas.

four-wheel steering (4WS)
A system whereby all four wheels are used to steer a vehicle.

FAP
The particulate filter *[Filtre à Particules]* is an element of filtration which reduces the polluting particulate emissions to the measurable limit.

It is a device made up of a porous silicon carbide surface, comprising channels which trap the particles with the passage of exhaust fumes. It is connected in its operation to a precatalysor placed upstream and temperature and pressure control sensors. The combustion of the particles which block it (natural from 550°C), is indeed impossible in the range of HDI engine use (150°-200°C). This combustion temperature is lowered to 450°C by injecting a compound containing cérine into the fuel (Eolys, developed by Rhodia).

G

gasoline
A liquid blend of hydrocarbons used as automotive fuel and processed from crude oil.

gear drive
A system of two or more gears, such as one that transmits power from the crankshaft or camshaft.

gear ratio
The speed relationship that exists between a driving (input) and a driven (output) gear. For example, a driving gear that revolves twice for each driven-gear revolution has a 2 to 1 (2:1) ratio.

generator
An ac or dc electrical-generating device that converts mechanical energy into electrical energy.

H

halogen lamp
A high-output, white-light lamp used as a headlight. Also known as halogen light.

hatchback
A passenger-car body style where a full opening hatch, including the rear window, lifts for access to the cargo storage area.

HC
An abbreviation for hydrocarbon.

HD
An abbreviation for heavy duty.

headlamps
The lights at the front of a vehicle used to illuminate the road ahead. Also known as headlights.

horsepower (HP)
A measurement of mechanical power or the rate at which work is done.

hybrid
A vehicle having two separate propulsion systems, such as a gasoline engine and an electric motor.

hydrocarbon (HC)
A compound containing hydrogen (H) and carbon (C), such as gasoline.

I

idle
The engine speed with no load and the accelerator pedal fully released.

ignition
The firing of a spark plug to ignite the air/fuel mixture in the combustion chamber.

indirect injection (IDI)
A system in which fuel is injected into a pre-chamber, where it is ignited before entering the main combustion chamber.

injector

1. The tube or nozzle through which fuel is injected into the intake airstream of the combustion chamber.

2. A term used for fuel injector.

ICE
An acronym for internal-combustion engine.

J

JAMA
An abbreviation for Japan Automobile Manufacturers Association.

K

KE-Jetronics
A continuous electronic fuel-injection system by Bosch that has been modified using a lambda oxygen sensor.

kickdown
A downshift to the gear below in an automatic transmission when the driver applies full throttle, as in overtaking and passing another vehicle.

L

lead
An element, Pb, often used as a body filler. To use lead as a body filler.

Short for tetraethyl lead, a compound formerly used to increase the octane rating of gasoline.

leakage current
A small amount of current that flows through insulation when a voltage is present and heats the insulation due to its resistance, resulting in a slight power loss.

Lean Burn
A Chrysler electronic engine control that appeared in the mid-1970s. It maintains precise control of the spark timing to allow a very lean mixture to burn properly, reducing emissions using an analog computer.

light-duty vehicle
Any motor vehicle rated at 8,500 pounds (3,856 kg) GVWR or less, having a curb weight of 6,000 pounds (2,722 kg) or less, and having a frontal area of 45 square feet (4.2 m^2) or less.

liquified natural gas (LNG)
Methane gas (CH4) that has been converted to a liquid by chilling.

liquified petroleum gas (LPG)
A predominately propane gas (C3H8) that contains some butane gas (C4H10) that is compressed to a liquid state.

M

manual transmission
A manually-shifted gearing device in the power train that allows variation on the relationship between engine speed and road speed.

manufacturer's code

1. An alpha-numerical code to identify a product.

2. A lettered marking code on tire sidewalls indicating tire manufacturer, plant, tire size, construction type and date of manufacture.

MCU
An abbreviation for microprocessor control unit.

mpg
An abbreviation for miles per gallon.

N

nitrogen oxide
Any chemical compound of nitrogen and oxygen (NOX) as a by-product of combustion that forms smog in the presence of sunlight.

nitrous oxide
A non-flammable, non-explosive gas (N_2O) used as an oxidizing agent with gasoline or methanol to increase the rate and efficiency of combustion, thereby increasing the horsepower.

O

O_2 sensor
A term used for oxygen sensor.

ozone
A molecule of oxygen, an unstable pale-blue gas (O_3), which is formed by exposure of O_2 to an electrical discharge. It has a penetrating or pungent odor and a strong oxidizing effect.

P

particulate
A solid matter, mainly soot from burned carbon, in an internal combustion engine's exhaust.

A form of solid air pollution such as microscopic solid or liquid matter that floats in the air.

particulate trap
An emission-control device in the exhaust system of a diesel engine which is used to capture particulates before they can enter the atmosphere.

parts per million (ppm)

1. A unit used to measure the amount of contamination in a substance, such as moisture in refrigerant.

2. A measurement of the emissions of a motor vehicle given as the number of parts of a particular chemical within one million parts of exhaust gas.

payload
The weight of the cargo that may be carried by a truck, determined by subtracting the curb weight of the vehicle and 150 pounds (68 kg) for each passenger from the gross vehicle-weight rating.

piston

1. An engine part that reciprocates in the cylinder and transfers the force of the expanding gases via the piston pin and connecting rod to the crankshaft.

2. A round, caliper component in a disc brake that is moved outward by fluid pressure to press the pads against each rotor face.

3. An aluminum or sintered-iron component of a drum brake inside a wheel cylinder that supports the cylinder cup.

4. The part of a compressor that is driven by a crankshaft to compress vapor.

planetary gear system
A gear set consisting of a central sun gear surrounded by two or more planet gears which are, in turn, meshed with a ring gear, used in overdrives and automatic transmissions.

platinum (Pt)
A rare, valuable metallic element which is highly resistant to corrosion, and is used as a catalytic agent in automotive catalytic converters of the oxidizing type.

pollutant
Any substance that adds to the pollution of the atmosphere.

Any substance in the exhaust gas from an engine or evaporating from the fuel tank or carburetor.

power steering
A power-assisted steering system that uses hydraulic pressure to increase the torque (turning effort) applied to the steering wheel by the driver.

R

rectifier
An electrical device used to convert alternating current to direct current.

S

stop & start
Particular management of the engine, aimed at reducing consumption, allowing to turn off the engine when the vehicle is stopped in neutral or when the brake is activated, and to start again instantaneously upon the driver's request.

Note: the system is active only provided that the engine is at its normal temperature of operation and that the battery charge is balanced or positive.

SUV
Concept of a multi-purpose vehicle, often top-of-the-range, of sporty appearance and "adventurous", generally with four driving wheels.

swirl
Like the tumble, the swirl is caused at the admission, with the goal of homogenizing the air-gasoline mixture in order to optimize the output of combustion. The swirl is the rotation movement of gases contained without the cylinder, around its axis.

Specialists characterize this movement by the “number of swirls”, which is the ratio of the speed of rotation of the gases to the engine speed.

T

TÜV
Administrative authority of the German Länders, entitled to carry out homologations and in particular those of the vehicles and automotive parts (comparable in this respect with the UTAC in France).

The TÜV thus issues marketing authorizations valid in Europe or in Germany only, according to the regulations to which it refers.

type mines
Name of the administrative reference under which, in France, the compliance in terms of motor-vehicle homologation was recorded. *Type Mines* was replaced by the NRE. As of 1997, it now only relates to commercial vehicles.

U

UTAC
Union Technique de l'Automobile et du Cycle (Technical Union for cars and bicycles). The UTAC is a French laboratory which carries out the tests intended for the homologation of vehicles. It also offers manufacturers its installations and competences in several fields: safety, acoustics, and environment, in particular.

V

valorization

This is the set of operations of reprocessing a vehicle out of use, recovering at the same time matter valorization – also called recycling – and energy valorization, i.e. use as a fuel or for energy production by any other means.

W

WAP

Wireless Application Protocol.

Appendices

Appendix 1. European regulation emissions for light vehicles

Gasoline vehicle (g / km)

Standard	Euro 2	Euro 3	Euro 4
Application year	1996	2000	2005
Cycle	91/441/EEC	MVEG	MVEG
CO	2.2	2.3	1.0
HC	0.275	0.2	0.1
NOx	0.225	0.15	0.08
HC + NOx	0.5	-	-

Diesel vehicle (g / km)

Standard	Euro 2	Euro 3	Euro 4
Application year	1996	2000	2005
Cycle	91/441/EEC	MVEG	MVEG
CO	1.0	0.64	0.5
HC	-	-	-
NOx	0.56	0.5	0.25
HC + NOx	0.7	0.56	0.3
Particulates	0.08	0.05	0.025

Appendix 2.a. Example of hybrid parallel transmission with flywheel storage

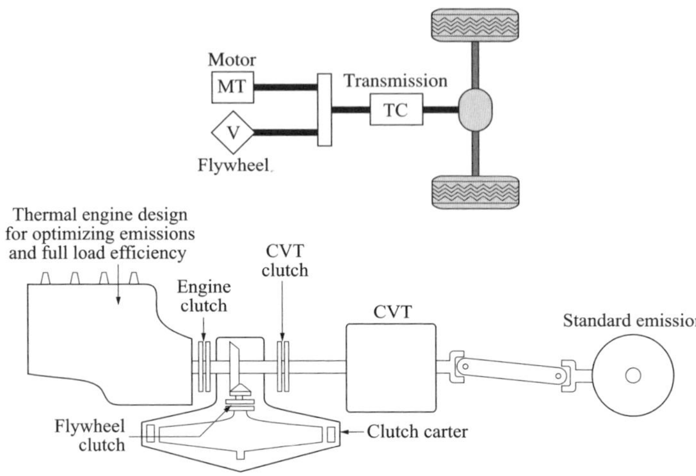

Appendix 2.b. Example of hybrid parallel transmission with oleo-pneumatic storage

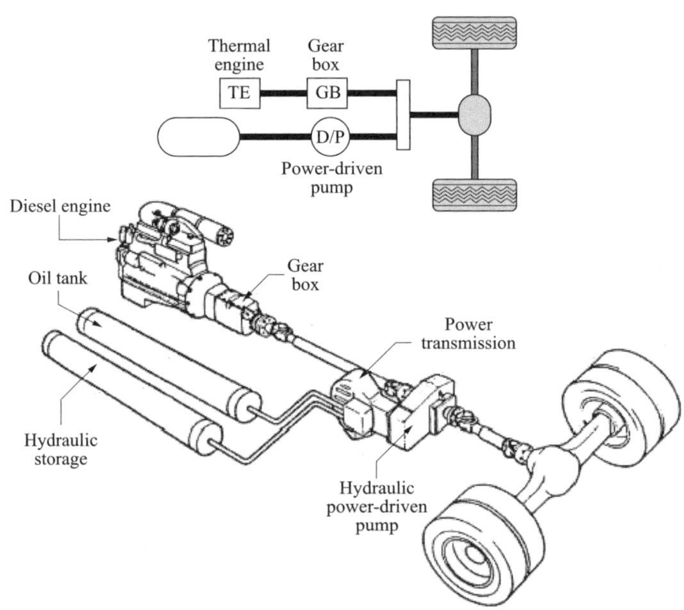

Appendix 3. Example of function allocation

		Functions					
Realizations	Type	SE	Boost	EM	ER	NC	SM
Toyota Prius 97, 00, 03	Dual	×	×	×			
Toyota Estima	Dual	×	×	×			×
Fiat Multipla	Dual	×	×	×	×	×	
Nissan Tino	Dual	×	×	×			
GM Precept	Dual	×	×	×			
ISAD concept	Parallel	×					
Toyota Crown	Parallel	×					
Honda Insight	Parallel	×	×				
Honda Civic	Parallel	×	×				
Ford Prodigy	Parallel	×	×				
Ford Escape	Dual	×	×	×			×
Chrysler ESX3	Parallel	×	×				
PSA Dynalto	Parallel	×	×				
PSA Dynactive	Parallel	×	×	×	×		
Renault Scenic	Parallel	×	×	×	×		
Renault Next	Parallel	–	×	×			×
Smart Hyper HEV	Parallel	×	×	×			
Chrysler Durango TTR	Parallel	–	×	×			×
Renault Kangoo	Series	×		×	×	×	
PSA Dynavolt	Series	×		×	×	×	
Renault/PSA VERT	Series	×		×	×	×	

Marketed (light grey)
Marketed in limited series (dark grey)
Prototype (white)

SE: Starting engine
EM: Electric mode
ER: Electric range
NC: Network charge
SM: Split motorization

Appendix 4. Toyota Prius engine

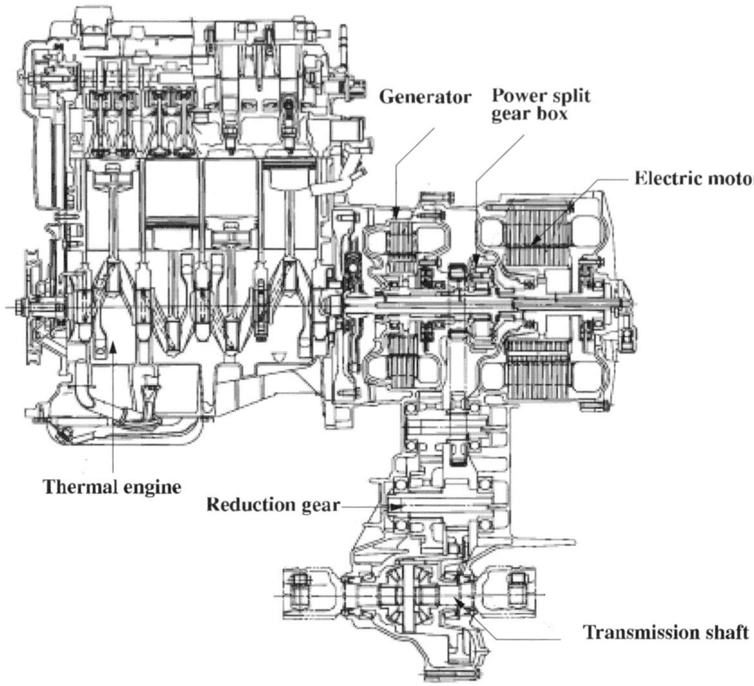

List of Authors

Thierry Alleau
Méaudre
France

François Badin
INRETS LTE
Lyon
France

Joseph Beretta
PSA Peugeot-Citroën
Vélizy
France

Cyriacus Bleijs
EDF–DTVE
Paris
France

Jean Bonal
ECRIN
Paris
France

Index

H

I

L

M

N

P

R

S